HISTORY

119th Infantry, 60th Brigade

30th Division

U.S.A.

Operations in Belgium and France

1917-1919

Coleman Berkley Conway

Written at the request of the Wilmington Chamber of Commerce and Published by that Organization in Honor of Col. John Van B. Metts and His Gallant Men and as a Contribution to American History

Originally Published by Wilmington Chamber of Commerce,

Wilmington, North Carolina, 1920.

Reformatted and Published by Jessica Duke,

Las Vegas, Nevada, 2011.

Book Layout and Cover Design by Jessica Duke.

Printed in the United States of America.

ISBN: 978-1-61393-021-2

TO THE PARENTS AND FRIENDS, AND, IN HONOR OF THOSE BRAVE AND NOBLE MEN OF THE 119TH INFANTRY, OF WHOM IT CAN TRULY BE SAID THAT THEY PERFORMED THEIR DUTY HONORABLY AND GLORIOUSLY.

This History of the 119th Infantry, 60th Brigade, 30th Division, U. S. A., was compiled by Captain C. B. Conway, of Danville, Va., and Lieutenant George A. Shuford, of Asheville, N. C. It was their effort to write only of FACTS, that the records of the deeds of true and brave men may be given. To them is due the thanks and appreciation of the officers and men of the Regiment.

Due to the loss of the list of the officers and the fact that a full and complete list cannot be published, the names of officers are not made a part of this record.

It is fitting to call attention to Major John Hall Manning, who as Operations Officer, rendered most excellent service; Major William S. Privott, Major Graham K. Hobbs, Major Robert S. Lamb and Captain Otho Robinson, who as Battalion Commanders, showed courage and ability of marked degree in fighting and caring for their commands in battle and at other times. Captain Jere Cooper, as Regimental Adjutant, and Captain Fred L. Black, as Supply Officer, were most efficient.

--JOHN VANB. METTS.

COLONEL JOHN VANB. METTS.

COLONEL JOHN VAN B. METTS.

ORGANIZATION OF 119th INFANTRY.

The 2nd North Carolina Infantry which went into Federal service June 16th, 1916, and spent six months on the Mexican Border was never mustered out upon its return, but proceeded to Camp Sevier, Greenville, South Carolina, August 1st, 1917. When the 30th Division was organized about Sept. 19th, 1917, the 2nd North Carolina Infantry was designated the 119th Infantry and assigned to the 60th Brigade.

A short while afterwards, the 59th Depot Brigade at Camp Sevier was broken up and the majority of the personnel of A, B, C, D, H, I, K, and L Companies of the 2nd Tennessee Infantry was absorbed into the 119th Infantry.

Later about 800 men (North Carolinians) were transferred from Camp Jackson, S. C., to the Regiment; and about 700 men were transferred from Camp Dodge and Camp Gordon and assigned to the 119th Infantry.

On or about September 5th, 1917, several Reserve Officers were transferred from Camp Lee, Va., to the 30th Division which was then under organization, and a number of these attached to the 2nd North Carolina Infantry. Some of these officers were later transferred to other Divisions, but those remaining in the Regiment were permanently assigned.

The personnel of the 119th Infantry, at the time it was ordered to France was as follows:

About 1,800 men from the State of North Carolina, 900 from Tennessee, and 700 from the States of Illinois, Indiana, Iowa, Kentucky, Minnesota, Ohio and Wisconsin.

TRAINING OF REGIMENT AT CAMP SEVIER, SOUTH CAROLINA.

Upon the completion of organization of the 119th Infantry, the Regiment underwent a most strenuous training preparatory to its final training received in France. A system of trenches was constructed by the 105th Engineers and these used extensively by the Regiment, in order that the men might become somewhat familiar with trench life, and their tactical use. A large target range was also constructed and on this the men were trained in the art of shooting accurately and rapidly, in order that they might protect their own lines in time to come. In the late fall of 1917, several British and French Instructors were attached to the 30th Division, and a few of these were sent to the 119th Infantry, who gave valuable instructions resulting from their own personal experience in France.

These instructors were specialists in their work, and this Regiment was fortunate in getting one for each branch of its training, namely: musketry and bayonet, gas defense, grenades, sniping and observation, stokes mortar, one pounder, and signal work. Beginning at about the 1st of January, 1918, detachments of officers from the 30th Division were sent each month to the School of Arms, at Fort Sill, Oklahoma, and the training for the officers continued until the Regiment was ready to depart for Overseas service.

Between January 1st, 1918, and April 1st, 1918, the Regimental Commander was in attendance at a Brigade and Field Officers' School at Fort Sam Houston, Texas. During his absence the Regiment was commanded by Col. F. LeJ. Parker, of the Regular Army, and his work for these three months greatly assisted in the building of the backbone of the Regiment. Many competitive drills and exercises were held during this time which assisted in establishing a keen spirit of competition, which the Regiment lived up to in later months.

Upon the return of Col. J. Van B. Metts from the School at Fort Sam Houston, Texas, he resumed command of the Regiment and carried on the training until the 30th Division was ordered to France.

ADVANCE SCHOOL DETACHMENT.

Previous to the embarkation of the 119th Infantry, orders were received to detail twelve (12) officers and several non-commissioned officers to be sent to France as an Advance School Detachment. The following officers were detailed:

MAJOR GEORGE K. FREEMAN,

MAJOR ROANE WARING,

CAPTAIN GRAHAM K. HOBBS,

1ST. LIEUT. CHARLES PICHETTI,

1ST. LIEUT. FRANK J. GILLIAM,

1ST. LIEUT. EDWARD L. GRAHAM,

1ST. LIEUT. CHESTER O. BELL,

1ST. LIEUT. GEORGE A. SHUFORD,

2ND. LIEUT. HOLCOMB ROGERS,

2ND. LIEUT. HARGROVE BELLAMY,

2ND. LIEUT. MARION R. WENTZ,

2ND. LIEUT. JAMES E. FAISON.

This Advance School Detachment, under the command of Major George K. Freeman, joined the general detachment of the 30th Division, which left Camp Sevier, April 30th, 1918, at 2:00 p. m., on a special train of the Southern Railroad. This detachment went direct to Camp Merritt, N. J., via Charlotte, Washington, Philadelphia, and Jersey City, arriving at Camp Merritt about mid-night of May 1st, 1918. While at this camp the N. C. O.'s were equipped for foreign service. On the morning of May 7th, 1918, the detachment boarded a train which carried the party to Hoboken, N. J., where it filed on board the transport George Washington, a converted German liner. Shortly after noon on May 8th, 1918, this liner slipped out of New York Harbor and started on its long journey to France. The convoy consisted of three (3) converted German vessels, the George Washington, the America and the DeKalb, formerly the Cruiser Crown Prince Eitel Frederich. The voyage was uneventful with the exception of one day spent in target practice. About 7:30 a. m., May 17th, 1918, seven (7) American Submarine Destroyers joined the convoy to see it safely into port. On the last day out at sea just at sunrise a submarine was sighted off the starboard side of the DeKalb. It at once submerged without firing. At 11:30 a. m., May 18th, 1918, the George Washington dropped anchor in the harbor of Brest, France. The Advance School Detachment debarked the following day, and marched to the Pantinzer Barracks, about 4 kilometers from Brest. On May 23rd, 1918, the party left Brest for Langres and arrived there May 25th, 1918. The entire detachment was here separated into different groups and commenced special training for which they were sent over. Some officers attended the School of the Line, the remainder the Specialist School. Upon the completion of the course the officers taking the Specialist and Line courses left Langres June 24th, 1918, and rejoined the Regiment which was stationed in the Racques Area, at Louches, on June 27th, 1918. Major George K. Freeman and

Major Roane Waring were transferred from the School of the Line to the Staff College before the completion of the course. These two officers were not reassigned to the Regiment.

EMBARKATION OF REGIMENT FOR OVERSEAS SERVICE.

Upon leaving the Training Camp at Sevier, S. C., the Regiment proceeded by special trains to Camp Merritt, N. J. Some of these trains moving over the P. & N. Railroad to Greenwood, S. C., and Seaboard from there to Richmond, Va., then to Washington, Baltimore, Philadelphia and Jersey City, N. J. Other trains moving over Southern to Charlotte, Danville, Richmond, etc. The entire Regiment was assembled at Camp Merritt, N. J., on the 8th of May, 1918, and remained there for a period of three days, being equipped for Over-Seas Service. While there, the organization was split up into three different detachments; one remaining there, one ordered to Philadelphia, and the other ordered to Boston.

Early on the morning of May 11th, Regimental Headquarters, Headquarters Company, Machine Gun Company, Supply Company, and Companies A, B, and C, under general command of Col. J. Van B. Metts. Companies A, B, and C, of the 1st Battalion and Battalion Headquarters, under command of Captain Otho Robinson, left Camp Merritt and proceeded by rail to the docks at Hoboken, N. J., arriving there about 9:30 a. m. No time was lost in transferring the troops to a large steamer, and in a few minutes landed on one of the large piers at New York used by Troop Transports. The detachment embarked the same afternoon about 4:30, on the British Steamship "Ascania", owned and operated by the Cunard Line.

The 2nd Battalion, consisting of Battalion Headquarters, Companies E, F, G, and H, under command of Captain Edgar H. Bain, left Camp Merritt, May 10th, and proceeded by rail to Philadelphia, Pa., arriving on the morning of the 11th. From the train the troops marched immediately to the docks, and at 2:00 p. m., embarked on the British Steamship "Haverford".

The third detachment, consisting of the 3rd Battalion Headquarters, Companies D, I, K, L, and M, under command of Major William S. Privott, left Camp Merritt, May 10th, and proceeded by rail to Boston, Mass., arriving there about 10:00 a. m., the 11th, and embarked that afternoon on the British Steamship "Laomadon".

VOYAGE.

The three British Transports, "Ascania", "Haverford" and "Laomadon" on departure from their respective ports made a direct sail for the coast of Canada, and on Wednesday, May 15th, saw each other in the Harbor at Halifax, Nova Scotia.

While in this harbor the convoy was formed, consisting of these three Transports and nine others. On the afternoon of May 16th, the convoy sailed for England. The "Ascania" was the flag-ship, and carried the Commodore of the convoy--Lieut. Commander Underwood, of the British Navy. The convoy was escorted by the British Cruiser "Cornwall".

Between the 17th and 21st the ocean was rather rough at times caused by heavy wind, but on the morning of the 22nd it was very calm, everyone was feeling better and in the best of spirits.

Friday, May 24th, the convoy reached the "danger zone", and at this time numerous "abandon ship drills" were being held. At daylight of the 25th several United States Submarine Destroyers were seen chasing all around our convoy, and remained as our best friends until the convoy landed.

About 11:30 p. m., May 26th, a German submarine was sighted within very close range, but it immediately submerged and was not seen again. The crafty Destroyers were on their job and dropped several "depth bombs" at the spot where the submarine had been seen. No disorder or confusion among the troops was caused during these crucial moments.

About 9:30 a. m., May 27th, 1918, the convoy filed through the Irish Sea and in a few minutes docked at Liverpool, England. All troops remained on board until 5:00 o'clock that afternoon when the Regiment debarked, and marched a short distance to the train which left at 10:00 p. m., for Dover, England's chief Channel Port for the exportation of troops. The train passed through London about midnight and arrived at Dover about 8:00 o'clock on the morning of the 28th.

Beginning at 11:00 a. m., on the same date, the Regiment moved by small detachments from Dover, England, across the English Channel to Calais, France, and by the afternoon of May 29th the entire Regiment was once more assembled in camp, located about one and one-half kilometers from Calais.

The Regiment remained in this camp for three days being equipped with gas respirators and ammunition. Here the United States rifles were exchanged for British rifles, as the 30th Division had been designated to serve with the British.

The night of arrival in this camp the Regiment experienced their first attack by the enemy in the form of an "air raid". In spite of the heaviness and nearness of the explosions, and the whining of the bomb fragments, the behavior of the troops in their first actual contact with the enemy was most excellent.

May 31st the organization moved by rail to Audrique, France, detrained there and marched to the Racques Area where the Regiment was billeted in the following towns: Regimental Headquarters, Headquarters Company, detachments of Supply Company and the 1st Battalion at Louches; 2nd Battalion and detachment of Supply Company at Landretheun; 3rd Battalion and detachment of Supply Company at Yeuse; Machine Gun Company at Nordt Ludlingham and Supply Company at Lostrat.

TRAINING PERIOD OVER-SEAS.

In this area the Regiment underwent a most intensive training schedule. Since it was to be affiliated with the British, their methods, the development of four years of painstaking study, was adopted. The 39th British Division, which was associated with the 30th (American) Division arranged to give assistance in training in an advisory capacity, by the attachment of British officers and N. C. O.'s to American units, and by the maintenance of close personal touch, also by the interchange of Staff Officers between the Staffs of the British formation and the corresponding Staffs of the 30th Division. The 119th Infantry was very fortunate in receiving a cadre of instructors from the 9th Black Watch and the 9th and 10th Gordon Highlanders, under the command of Brig. Gen. N. S. Hubbard, 118th Infantry Brigade (British) and under the direct supervision of Col. Cruickshenk, Lt.-Col. Sworder, Lt.-Col. Lord Gordon and Lt.-Col. Anderson. The training was divided into two phases, i. e.: A and B. While in this area, phase A was mastered. This training consisted of 36 hours of instruction for the first week, 25 hours for the second week, 26 hours for the third week, and 20 hours for the fourth week. Realizing that every minute must be taken advantage of, the men worked most industriously, with a spirit at which our Allies marvelled. The training of specialists received the most careful attention, as did the training of the Infantry platoons. It was due, in a great way, to this thorough training that the later phenomenal success of the Regiment in actual battle is attributed. Time was also given to March Discipline and Trench Warfare.

As another means of taking advantage of the experience gained by the British, officers and non-commissioned officers were sent to various schools to acquire technical knowledge of Automatic Rifles, Bayonet Fighting, Grenades, Gas Defense, etc. It is well to know that the average standard set at these schools by the 119th Infantry officers and N. C. O.'s was exceptionally high. Also, officers and N. C. O.'s were sent to the School Sector directly east of Ypres for practical experience under actual battle conditions before receiving the responsibility of leading their men under enemy fire. They were most courteously received by the units of the 49th British Division, occupying the line and were given valuable information and instruction while there. Some of these officers and N. C. O.'s were permitted to go on patrol, but chiefly they acquainted themselves with trench life and warfare.

The Machine Gun Company with the Machine Gun Companies of all the Regiments in the Division, were assembled at Nordt Ludlingham, where they were instructed in the latest methods of Machine Gun Warfare which the British had found so successful.

While in this area authority was received from G. H. Q., for the enlargement of the Regimental Staff by the allowance of an Operation and Intelligence Officer. To fill this very important office, Captain, now Major John Hall Manning, then the Regimental Adjutant, was appointed. Lieut. Jere Cooper was promoted to the rank of Captain and transferred to Regimental Headquarters as Adjutant. Captain Benjamin West on June 4th was also transferred to Regimental Headquarters as Personnel Adjutant. Upon the completion of the above changes, the Regimental Staff consisted of:

COL. J. VAN B. METTS,

LIEUT.-COL. CYRUS M. FAIRCLOTH,

CAPT. JOHN HALL MANNING,

CAPT. BENJAMIN WEST,

CAPT. JERE COOPER,

CAPT. ARMSTEAD K. TAYLOE,

CHAPLAIN JOHN M. ROBESON.

The 119th Infantry Regiment was inspected almost immediately upon its arrival in this area by Major Gen. Herbert Plummer, Commander of the 2nd British Army. A week later Field Marshall Sir Douglas Haig, Commander of the British Armies in France, honored this Regiment by giving it a most rigid inspection. Both of these officers expressed great delight in the progress of the training and the wonderful physique of the men. On or about the 30th day of June, Gen. John J. Pershing, Commander in Chief of the American Armies in France, inspected the Regiment. He was pleased with the fitness of the troops and especially the high standard of the personnel, expressing himself that he had not seen a better body of men in the American Expeditionary Forces.

About the 15th of June, information was received that an attack by the Germans was anticipated in the sector immediately east of Cassel. In anticipation of this plan of attack arrangements were perfected whereby the 30th Division was to take up a position in reserve, west of Cassel. The 119th Infantry was designated to be the Divisional Reserve which, as expressed to the Commanding Officer by Gen. Faison, was a very important position in the scheme of defense. The Regiment was preparing for this move with great enthusiasm. This position, however, was never occupied. The orders were unexpectedly changed and this Regiment with the other units of the Division, on the 2nd day of July, 1918, proceeded by marching to the Roozendaal and Road Camp Areas, near Watou, Belgium.

The Regiment formed on the Zaufques-Louches-Yeuse Road with the head of the column at the road intersection 3A, 39, British Map Hazebrouk, at 9:45 a. m., in the following order: Headquarters Company, Detachment Supply Company, 1st Battalion, 3rd Battalion, and 2nd Battalion. The Sanitary Detachments accompanied their respective Battalions. The First and Second Line Transports followed their respective Battalions in the following order: Rolling Kitchens, G. S. Wagons (Ration), G. S. Wagons (Baggage), Water Carts, Limbers. The following disposition of equipment was made: 120 rounds of service S. A. A., one cooked meal and full field equipment, including two blankets, less overcoats, which were carried on the wagons, were carried by each man. With this equipment and in the above order the 119th Infantry marched to the Foret D'Eperlecques, a distance of 12 kilometers, where they bivouaced the night of July 2-3, 1918. On the morning of July 3rd, the Regiment marched to Rubrouck, forming at 6:15 a. m., on the Watten-Welverdinghe-Voikermakhove-Herzeele Road in the following order: Headquarters Company, Detachment Supply Company, 3rd Battalion, 2nd Battalion, 1st Battalion, Sanitary Troops with their respective Battalions, where it bivouacked the night of July 3-4, 1918, in a pasture.

The distance marched was about the same as on the day before. On July 4th, 1918, contrary to expectations, the Regiment again formed on the Bollezeele-Rubrouck Road in the following order: Headquarters Company, Detachment Supply Company, 2nd Battalion, 1st Battalion and 3rd Battalion, with Sanitary Troops with their respective Battalions at 7:20 a. m., and marched to Herzeele, where it bivouacked for the night of July 4-5, 1918. Here it was that the men of this Regiment first heard distinctly the firing of guns on the famous Flanders front. It was the subject of much discussion especially since it was the birthday of the Independence of America. On the morning of July 5, 1918, at 6:00 a. m., the Regiment formed on the Herzeele-Watou Road, to march to the Roosendall and Road Camp Areas in the order of march as of the first day. It reached its destination shortly after noon and went into camp two kilometers southeast of Watou, Belgium. It is quite interesting to note that this Regiment was the first American unit to enter Belgium. Regimental Headquarters and Headquarters Company were located at Bayley Camp, 1st Battalion was located just across the road, 2nd Battalion at Gabb Camp, 3rd Battalion at Chase Camp, and the Machine Gun Company in a pasture only a short distance away. The Supply Company, less the detachments which were attached to the different Battalions formed the Sixtieth Brigade Dump and was located at Dragon Camp.

It must be understood that the Supply Company of this Regiment and the Supply Company of the 120th Infantry Regiment were consolidated forming the Sixtieth Brigade Dump, each Battalion was then allotted a detachment under the command of a Battalion Supply Officer. It was the duty of the Brigade Dump to issue rations and equipment for a Battalion directly to the respective Battalion Supply Officer, who in turn issued the same directly to the Company of their Battalion. Also in this detachment was included a Battalion Transport Officer who had direct charge of the Battalion Transport. The feeding of the men in the line was entirely in the hands of these two officers, for as soon as the Brigade Dump had turned over the food to the Battalion Supply Officer he was relieved of further responsibility. Then the Transport Officer after the rations had been prepared for consumption collected them and loaded them on the transports, carrying it forward to the men in the trenches. This Regiment was fortunate in having conscientious and fearless officers to fill these important positions and they deserve the highest praise for the manner in which they performed their duties. This method of supply was continued throughout the operations of the Regiment.

After resting up from the four-day march, the Regiment resumed the training schedule with renewed energy. The front line at its closest point was about 12 kilometers distance, consequently the firing of artillery could be heard at all times, and even the bursting of shells, for on several occasions the town of Watou, where Division Headquarters was located was shelled. The shells passing directly over the Camp. Here also the men of the Regiment became thoroughly accustomed to enemy air raids. Practically every clear night there were numbers of German planes overhead bombing important cross-roads and camps. The transports carrying rations forward to the men in the line also presented a favorite target.

It was while stationed in this area that Major Gen. Pinny, of the British Army, addressed the officers and non-commissioned officers of the Regiment. He was very enthusiastic on the condition of the men, especially the standard of morale, which was extraordinarily high. There was one remark in his speech which clearly bespoke the conditions and thinking of our gallant allies before our arrival. Gen. Pinny said in substance: "An attack by the enemy is anticipated and in all probability his advance will be on this front. It is my sincere hope that he attacks while you are here." The attack never materialized.

To better understand the subsequent work of the Regiment it is here necessary to explain somewhat in detail the former training which was conducted under the heading of Phase B training. This consisted of one Battalion at a time going into the line and being associated with the British units then occupying the position and also to strengthen the line as it was very lightly held. The command and the responsibility for this sector was retained by the British while the Americans were given every opportunity to accustom themselves with trench life and conditions. This method of training was to continue until the entire outfit had been able to take advantage of the opportunity. Phase B training was commenced in the Regiment by the following order, July 16, 1918:

HEADQUARTERS, 119TH INFANTRY
American Expeditionary Forces
FRANCE, JULY 15, 1918

SECRET

FIELD ORDER No. 9

Reference Map:

--Sheet 27, N. E.

--Edition 3-A (Local)

--Sheet 28, N. W.

1. This Regiment, in conjunction with the remainder of the 60th Infantry Brigade, will commence training under Phase "B" scheme, July 16, 1918.

2. (*a*) The 1st Bn., less Co. B, will march on Zwynland Brewery at 7:45 p. m., July 15, 1918.

(*b*) The 2nd Bn., will march on Zwynland Brewery at time to be designated on July 16, 1918.

(*c*) The 3rd Bn., and Hd. Co., will remain in their present camps.

(*d*) The M. G. Co., will receive instructions from the 30th American or 33rd British Divisions.

(*e*) Battalions will be subject to the orders of the British Units to which attached upon arrival in the Blue Line.

3. (*a*) The 1st Bn., less Co. B, and the 2nd Bn., will march via Cross Roads, L. 1, c. 9. 1.--Cross Roads L. 14. c. 4. 3.--L. 14. b. 1. O,--Road Intersection L. 21. c. 95. 95.--thence along Main Abeele-Poperinghe Road to Zwynland Brewery, where guides from 33rd British Division will be met to conduct the Bn., into bivouac along the East Poperinghe Line.

(*b*) Column will march with 100 yd. intervals between Companies; 100 yd. interval between unit and its transport; and 25 yd. intervals between each section of six vehicles.

4. (*a*) Only the necessary transportation will be carried.

(*b*) Hospital Corps Detachments will accompany their respective Battalions.

(*c*) Prescribed amount of ammunition will be carried.

(*d*) One day's cooked ration in addition to the iron ration will be carried in the haversack of each officer and man.

(*e*) Shelter halves, poles, pins and blankets will be left under guard of 1 N. C. O., and 6 Pvts., who will be attached to the Hq. Co., 119th Inf., for rations. Equipment of officers will be limited to that carried on their persons.

5. Regimental Headquarters will remain at Bayly Camp (K. 12, d. 65. 75).

6. Acknowledge.

J. VAN B. METTS,
Colonel, 119th Infantry.

While undergoing Phase B training the following plan was adopted. This order was issued so that each unit would know exactly what to do in case of an enemy advance which, as stated, was anticipated on the immediate front.

HEADQUARTERS, 119TH INFANTRY
American Expeditionary Forces
FRANCE, JULY 17, 1918

SECRET

Instructions for occupying East Poperinghe Line

Reference Maps:--9145, Sheets 27 and 28.

1. During the Phase "B" training in case of emergency, the 60th Brigade will hold the Blue Line. Brigade P. C. will be established at Road Camp F. 25. d. 4. 1.

2. This Regiment, less 1st Bn., which is under orders of the 33rd Division (British) in front line and M. G. Co., will occupy the entire Brigade Sector of the Blue Line upon receipt of message "Occupy East Poperinghe Line".

3. (*a-1*) The 2nd Bn., will cover the right sub-sector forward line with 2 Cos., extending from G. 14. d. 6. 6. and G. 14. d. 85. 55. (right) to G. 9. c. 5. 2. and G. 15. a. 7. 0. (left) with Bn., P. C. at G. S. a. 1. 3. (Elizabeth Chateau.)

(*a-2*) Support line with 2 Cos. from G. 14. a. 55. 00 (right) to G. S. d. 75. 85. (left).

(*b-1*) The 3rd Bn., during the absence from this area of the 1st and 3rd Bns. 120th Inf. will march from the present camp by the most convenient route to the Blue Line and will cover the left subsector forward line with 3 Cos., extending from G. 9. c. 5. 2. and G. 15. a. 7. 0. (right) to G. 4. c. 3. 1. and G. 4. c. 8. 1. (left) with Bn. P. C. at G. 2. d. 4. 1.

(*b-2*) Support line with 1 Co. from G. 9. a. 3. 3. (right) to G. 9. b. 2. 8, (left).

(*c*) Each Bn. and Co. position will be covered and occupied as shown on Map 9145, as marked and issued by 30th Division (American).

(*d*) 1 Bn. of the 59th Brigade will be in support in this sector with P. C. at G. 8. a. 1. 3.

(*e*) 174th Brigade F. A., 39th Divisional Artillery (British) will cover the East Poperinghe Line. S. O. S. lines will be issued to those concerned.

(*f*) Liaison will be maintained with the 6th Division (British) on the right flank Brigade Sector by the 2nd Bn., and with the 59th Brigade on the left flank of the Brigade Sector by the 3rd Bn. Liaison will be maintained between units of the Regiment at all times.

(*g*) Flank Companies in the Brigade Sector will each send forward a platoon to the general line Linda Goed Farm--Road Junction in G. 5. d. 0. 2.--Oosthoek System (A. 23. d.) for the purpose of maintaining liaison between our own front line troops and the British Bn. in the Ouderdom-Brandhoek System. They would also cover the withdrawal of British units from advanced positions.

(*h*) Sector Commandant (Hq. G. 8. d. 7. 6.) is in charge of ammunition dumps S. A. A., Stokes ammunition and grenades:

No. 1. G. 8. c. 8. 4. for 119th Inf.

No. 2. G. 8. d. 7. 6. for 120th Inf. (normally under this scheme for 3rd Bn., 119th Inf.)

(*i*) Units will proceed equipped with arms, ammunition, pack, consisting of overcoat (one blanket may be carried in lieu of overcoat), haversack, containing mess equipment, toilet articles, iron rations, two pairs socks, slicker, rolled on outside of pack, helmet worn, box respirator in alert position, bayonet on pack.

4. Wagon trains will continue Hqrs. in present location.

5. Battalions will report immediately upon completion of the occupation of their position.

6. Regimental Hqrs. will move to Zwynland Brewery.

Acknowledge.

J. VAN B. METTS,
Colonel, 119th Infantry.

In compliance with the above order the 1st Battalion under command of Capt. Otho Robinson, moved to the Blue Line (reserve position east of Poperinghe) on the night of July 15th, 1918. Company "A" occupied the right front and right support with two (2) platoons in the front line position and two in support. Company "C" occupied exactly the same relative position on the left. Company "D" was held in reserve. On the night of July 16-17 the Battalion moved to the front line position in the right subsector of the Canal sector in the Ypres salient. The disposition of troops in the sector were as follows: "A" Company occupied the front line in conjunction with "C" and "B" Companies of the Argyle and Southern Highlanders with their right resting on Kruisstraathock-Voormezeel Road. To the right of "A" Company and occupying the front line "D" Company was stationed with two platoons in the front line and two in close support with the Queens Royal Rifles. "C" Company occupied support positions, "B" Company received orders to return to Louches, France, by rail to complete its course on the target range.

On the same night the 2nd Battalion under command of Capt. Graham Hobbs, took over the position formerly occupied by the 1st Battalion in the Blue Line.

For the first six days that this Battalion was in the line it held the position with the British, first by having one English soldier stationed with one American. After confidence was thus gained the separate units were grouped together and a British platoon was placed on either side of an American platoon. On the 23rd day of July, after the Battalion had been in line for a period of six days the British withdrew to the Yellow Line or the Regimental Reserve Line and the 1st Battalion held the front line position alone until relieved by the 2nd Battalion on the night of July 24-25, 1918.

This sector was known as a quiet one and especially suitable for the instruction of new troops but at this particular time it was quite the contrary. As has been stated, an attack was anticipated, so the British artillery was exceptionally active firing on all points where troop movements had been noticed and on important roads and crossings. The artillery's activity on our part consequently called for retaliation from the enemy which made the sector one of the hottest on the front. Due to this very heavy shelling the casualties numbered more than was expected.

The 2nd Battalion was disposed of as follows: "F" Company occupied the right front sector with two platoons in the front line and two platoons in close support. "E" Company with four platoons occupied the front line on the left sector. "G" Company with four platoons in close support. "H" Company in reserve. "F" and "H" Companies were with the Queen's Royal Rifles, while "E" and "G" Companies were with the 5th Scotch Rifles. During the period of occupation there were a few changes made in this disposition by means of inter-Battalion reliefs. The training was the same as stated for the 1st Battalion.

The 3rd Battalion occupied the Blue Line during the period that the 2nd Battalion was in the forward position then relieved the 2nd Battalion in the Canal sector on the night of August 1-2, 1918. A detachment from the Signal Platoon from Headquarters Company under command of 1st Lieut. Nathaniel Hannau, accompanied the Battalion into line.

The Machine Gun Company was under control of the 33rd British Division and received practical instructions in the same manner as did the line companies of the Regiment.

While the 3rd Battalion was occupying the front line position, the 1st Battalion held the reserve trench or Blue Line going in on the night of August 1-2, and remaining until it was withdrawn Sunday morning, August 10th, 1918.

With the withdrawal from the line of this Battalion Phase B of the scheme of instruction was completed for this Regiment and it was now deemed capable of taking over and occupying a place in the firing line alone.

From the 10th of August, 1918, until the 15th day of August, 1918, the Regiment trained around Watou in a great way correcting errors which had been noted during the tour in line. On the 15th day of August, the

Commanding Officer of the Regiment received orders that the Regiment should proceed that night and occupy the Blue Line in the Canal sector, which it did. Regimental Headquarters was established at the Elizabeth Chateau on the eastern edge of the town of Poperinghe.

HEADQUARTERS, 119TH INFANTRY
American Expeditionary Forces
FRANCE, AUGUST 15, 1918

SECRET

FIELD ORDER No. 14

Reference:--Sheets 27 and 28

1. The 119th Infantry Regiment will relieve the 19th British Brigade in the Reserve Brigade Area of the 33rd British Division the night of 15-16 August, 1918, in accordance with attached table. All moves will be by platoon at 100 yds. intervals. Battalions will send advance parties of one officer, per Bn. Hqrs. and one officer per Company to report to Hqrs. of the British Bns., to be relieved at 5 p. m., today. The Hqrs. Company will send one officer and one N. C. O. per platoon to the 19th British Brigade Hqrs., at 5 p. m., today.

2. The 19th British Brigade is providing guides at the scale of:

1 per Bn. Hqrs.

1 per Co. Hqrs.

1 per platoon.

3. Battalions of the 19th British Infantry Brigade are handing over to relieving American Battalions, all defense schemes, dumps of ammunition, grenades, all area stores, and all 1-10.000 maps.

Instructions regarding the provision of the sector commandants of the Green and Yellow Lines are being handed over by Bn. Hqrs. of the Bn. in the Green Line and the left Battalion in the Yellow Line.

4. The transports will remain in their present camps. Guards will be provided for property at the rate of 1 N. C. O. and 6 Pvts. per Bn., to be retained by the transport.

5. Completion of relief will be reported to Regimental Headquarters. Regimental Headquarters will close at Bayly Camp and reopen at Elizabeth Chateau (G. 8. a. 2. 2.) at 9:00 p. m., tonight, August 15.

6. Orders for the movement of the Regimental Machine Gun Company will be issued separately.

7. On the night of 16-17 August, 1918, the 119th Inf. will relieve the 98th British Brigade in the right sub-sector of the 33rd British Divisional Front.

8. Advance parties as usual will report at Regimental Hqrs., at 6:00 p. m., today and proceed to the light railway from Remy Siding to Brisbane where they will be met by guides from the 98th British Brigade and taken into the line.

Regimental Hqrs	1 Officer	4 Signallers
Each Bn. Hqrs	1 Officer	2 Signallers
Each Co. Hqrs	1 Officer	
Each Platoon	1 N. C. O	
Each L. G. Team of Bns., going into line	1 other Ranks	
Stokes Mortar	1 Officer	

These men will make themselves thoroughly acquainted with the area to be taken over on the night of 16-17 August, and will take over from the British units, defense schemes, maps, dumps of ammunition, grenades, etc., and all trench stores.

9. The relief on the 16-17 August will be carried out as follows:

(*a*) 1st Bn. will take over the right section relieving the 2nd Bn., A. and S. H., with Hqrs. at H. 30. a. 5. 9.

(*b*) 2nd Bn. will take over the left section relieving the 4th Kings, with Hqrs. at H. 24. c. 6. 1.

(*c*) 3rd Bn. will take over the reserve relieving the 1st Middlesex, with Hqrs. at H. 22. a. 4. 4.

10. Further orders regarding this relief will be issued later.

11. Units of the 19th British Brigade are sending advance parties to take over camps being evacuated by American units. These parties will arrive at Bn. Hqrs. of the American units concerned at 5:00 p. m.

By Order of COLONEL METTS,

JOHN H. MANNING,
Captain O. and I. O.

OCCUPATION AND OPERATIONS IN CANAL SECTOR UP TO, AND INCLUDING SEPTEMBER 4th, 1918.

The Canal sector was the general line extending from immediately southeast of Ypres, about two miles southwest to Elzenwalla, inclusive, on both sides of the Ypres-Commines Canal and the country on this immediate front was very low and wet, thus causing many hardships upon the troops occupying it. To the right of the Canal sector stood Mount Kemmel, from the top of which the Germans had a dominating view of the entire sector, thus causing camouflaged screens to be erected on all main roads leading towards the front, and making it very difficult to move about during the day. On the left of this sector was the remains of Ypres, after the great battle the British fought in July, 1916, when the Germans used gas for the first time.

The subdivisions of the Canal sector, such as support, reserve, etc., were represented by different lines of trenches as follows:

The actual front line was known as G. H. Q. 1, and the second in close support as G. H. Q. 2 line. Support position was the Brown Line, and the reserve Yellow and Green Lines.

This front when taken over by the 30th American Division was occupied by the 33rd British Division, and the 89th British Brigade under the command of Brigadier General Baird, was holding the front line. This Brigade consisted of three battalions as follows:

2nd Argyle and Southern Highlanders.

4th Kings' Royal Rifles,

1st Middlesex.

On August 15th, 1918, the following order was issued from Headquarters 119th Infantry:

HEADQUARTERS, 119TH INFANTRY
American Expeditionary Forces
FRANCE, AUGUST 15, 1918

SECRET

FIELD ORDER No. 15

Reference Map:--Sheet 28

1. The 119th Infantry Regiment (less 3rd Bn.) will relieve the 98th British Brigade in the right sub-sector of the Canal sector on the night of August 16th and 17th.

The 3rd Battalion, 119th Infantry Regiment will remain in its present position in the Green Line and will be in Regimental reserve.

2. The Relief will be carried out as follows:

(*a*) The two British Companies in the front line and outposts in each forward Battalion area will be relieved by one American Company.

(*b*) The two British Companies in support and reserve in each forward Battalion area will be relieved by one American Company.

(*c*) The remaining two Companies of the 1st and 2nd Battalions will relieve the two British Companies of the reserve Battalion of the 98th British Brigade in their respective areas in and about the Brown Line.

On completion of the relief the 1st and 2nd Battalions, 119th Infantry Regiment, will each be disposed with one Company in the front line and outposts, one Company in support and two Companies in reserve in and about the Brown Line.

3. All moves will be by platoons at 100 yards.

4. The 98th British Brigade is providing guides to meet the 1st and 2nd Battalions, 119th Infantry Regiment, at Road Junction H. 22. a. 35. 30. at the rate of:

1 per Battalion Headquarters,

1 per Company Headquarters,

1 per Platoon.

5. Units will take over from units of the 98th British Brigade to be relieved all defense schemes, maps, information about the enemy, and details of work in hand proposed, also trench stores.

6. 98th British Brigade are leaving officers and N. C. O.'s. to remain with incoming American units for 24 hours, as follows:

Regimental Headquarters	1 Officer,
Each Forward Bn. Headquarters	1 Officer and Bn. I. O.,
Each Company Headquarters	1 Officer and 1 N. C. O.,
Stokes Mortar Platoon	1 Officer.

7. All other details of the relief will be arranged between Commanding Officers concerned. O. C. of 1st and 2nd Battalions, 119th Infantry Regiment will be expected at the Headquarters of 2nd Argyle and Southern Highlanders, and the 4th Kings, respectively, at 10:00 a. m., August 16th, and should both visit Headquarters of 1st Middlesex afterwards.

8. The 120th Infantry Regiment is moving two Battalions to the Yellow Line on the night, August 16-17, which will take over as follows:

1st Battalion, 120th Inf. Reg., from 2nd Bn. 119th Reg., in the left sector of the Yellow Line.

2nd Battalion, 120th Inf. Reg., from 1st Bn. 119th Reg., in the right sector of the Yellow Line.

1st and 2nd Battalions, 120th Infantry Regiment are each sending advance parties of 1 officer per Battalion Headquarters, 1 officer per Company and 1 N. C. O. per Platoon, to report to the 1st and 2nd Battalion Hqrs., 119th Infantry Regiment, respectively, at 3:00 p. m., August 16th. These parties will take over all dumps of ammunition, and all trench and area stores in the Yellow Line.

9. 119th Regiment Machine Gun Company will march from its present camp at 3:00 p. m., August 16th, to Chinese camp, G. 3. d. 6. 4,

and will come under the orders O. C. 33rd British Battalion, M. G. C., on arrival.

119th Regiment M. G. C., will send advance party of 1 officer and 2 N. C. O.'s to Chinese camp to take over accommodations.

119th Regiment M. G. Co., will move into the right sub-sector of the line on the night of August 17-18. Orders for this move are being issued by the O. C. 33rd British Battalion M. G. Co.

10. Completion of relief will be reported to Headquarters, 98th British Brigade.

11. Command of the right sub-sector of the Canal sector will pass to Colonel commanding 119th Infantry Regiment on completion of the relief on night of August 16-17.

12. Headquarters 119th Infantry Regiment will close at Elizabeth Chateau and re-open at Assan Farm, H. 22. a. 2. 6. at 9:00 p. m., August 16th.

J. VAN B. METTS,

Colonel, 119th Infantry.

13. Acknowledge.

14. Distribution of copies:

1--C. G., 30th Division.

2--C. G., 60th Brigade.

3--1st Battalion.

4--2nd Battalion.

5--3rd Battalion.

6--Headquarters Company.

7--Machine Gun Company.

8--Supply Officer.

9--33rd British Division.

10--98th British Brigade.

11--33rd British Bn., M. G. C.

12--120th American Regiment.

13--Surgeon.

14--War Diary.

15--Files.

The following morning, August 16th, orders were issued for moving the 3rd Battalion, 119th Infantry Regiment from the Green Line to the right sub-sector of the Yellow Line.

Upon moving into the front line on night of August 16-17, the 119th Infantry Regiment took over from the 98th British Brigade in the following order:

The 1st Battalion, commanded by Capt. Otho Robinson, occupied the right Regimental front. Company "A" relieving two companies of the Argyle and Southern Highlanders in the front line and Company "B", relieving two companies of the Argyle and Southern Highlanders in close support. One platoon of Company "B" occupied a small trench in Scottish Woods, acting as a counter-attack platoon. Companies "C" and "D" occupied support trenches in the Brown Line near Smythe Farm, relieving two companies of the 1st Middlesex.

The 2nd Battalion, commanded by Capt. Graham K. Hobbs, occupied the left Regimental front. Company "H" relieving two companies of the 4th Kings in the front line and two isolated trenches on the left front, about 200 yards from the main trench line. Company "F" taking over from two companies of the 4th Kings in the G. H. Q. 2 line as a close support. Companies "E" and "G" occupied support trenches in the Brown Line near Assam Farm.

Machine Gun Company upon orders from British Battalion M. G. O., moved its headquarters to Smythe Farm with that of Company "D", 1st Battalion. Practically all of the Gun positions of the M. G. Co., were in the G. H. Q. 2 line.

Stokes Mortar and one pounder platoons were located in the Ravine. 1st and 2nd Battalion Headquarters were located between Belgian Chateau and Segard Wood. Regimental Headquarters was located at Assam Farm.

When the 30th Division left the States for Overseas service the Divisional Artillery was left behind and did not come over until about one month later. Upon its arrival in France, it was assigned to duty with the 33rd American Division in the Verdun sector.

When the 33rd British Division was relieved on August 16-17, the 98th British Artillery Brigade, commanded by Colonel Butler, remained in the line supporting the right sub-sector until about August 25th, when it was relieved by the -- British Artillery Brigade, commanded by Colonel Ladd, with headquarters at Assan Farm and was still in the line when the 30th Division was relieved.

On the night of August 21-22, 1918, the 3rd Battalion, commanded by Major William S. Privott, which was then in the Yellow Line relieved the 1st Battalion, 119th Infantry Regiment, then in right Regimental front. The 1st Battalion upon being relieved withdrew to the same position vacated by the 3rd Battalion. The disposition of the 3rd Battalion was as follows:

Company "M" took over right Regimental front. Company "K" had one platoon in Scottish Wood with the remainder of the Company in the Domino Trench and the Ravine. Company "L" in the Brown Line near Ottawa Farm. Company "I" in the Brown Line near Smythe Farm.

On the night of the 22-23 of August, the Machine Gun Company was relieved by Company "B", 115th Machine Gun Battalion. Upon the completion of the relief, the Machine Gun Company of 119th Infantry Regiment, proceeded to Chinese Camp.

Following is the defense scheme of the right sub-sector occupied by the 119th Infantry Regiment (American):

NAHU, AUGUST 28, 1918

SECRET

COPY No. 1

Defense Scheme--Right Sub-Sector--Canal Sector

Reference Maps:--

Sheet 28, N. W. 3--} 1-10,000

--Sheet 28, N. W. 4--} 1-10,000

1. The 30th American Division holds the general front line of the Canal sector which extends from the Railway, incl at H. 36. c. 9. 5. on the right, by Gunners Lodge (I. 26. b.) to the Railway, incl. at I. 21. d. 3. 6. just South of Zillebeke Lake on the left.

2. This Regiment holds the right sub-sector of the 30th American Division (Canal) sector with two (2) Bns., in the front line and one (1) Bn., in reserve.

3. A Regiment of the 27th American Division is on our right.

A Regiment of the 30th American Division is on our left.

4. BOUNDARIES--DIVISION--

Southern Boundary:

The Elzenwalle-Vijverhoek Railway, incl from H. 36. c. 9. 5. to H. 29. c. 4. 6., thence to H. 29. a. 3. 0.--H. 22. c. 0. 0.--H. 19. b. c. 5. 6.--G. 13. Central.--L. 17. b. 7. 0.

Northern Boundary:

The Railway, incl., from I. 21. d. 3. 6. to I. 13. d. 0. 4.--thence H. 18. b. 9. 0.--H. 12. c. 4. 0.--H. 10. b. 9. 0.--along Railway to G. 2. d. 3. 0.

BOUNDARIES--REGIMENTAL--

Southern Boundary:

The Southern Divisional Boundary.

Northern Boundary:

I. 25. d. 35. 10.--along track running N. W. to I. 25. a. 7. 0.--G. H. Q. Line at I. 25. a. 25. 60.--thence along old communication trench (Beliegoed Avenue), excl to H. 24. d. 3. 1.--H. 24. c. 9. 7.--H. 23. b. 35. 45.--H. 16. a. 0. 0.--Inn Corner.

BOUNDARIES--RIGHT BATTALION--

Southern Boundary:

The Southern Divisional Boundary.

Northern Boundary:

The Voormezeele-Kruistraathoek Road from I. 31. a. 0. 2., H. 30 d. 55. 05, excl. thence old Communication Trench to H. 30. a. 78. 30, incl. H. 30. a. 8. 5., thence track to H. 30. a. 35. 90.

BOUNDARIES--LEFT BATTALION--

Southern Boundary:

The Voormezeele-Kruistraathoek Road from I. 31. a. 0. 2., H. 30 d. 55. 05, excl. thence old Communication Trench to H. 30. a. 78. 30, incl. H. 30. a. 8. 5., thence track to H. 30. a. 35. 90.

Northern Boundary:

The Northern Regimental Boundary.

5. The lines of defense consist of:

1.--The front line consists of a number of unconnected advance posts which are changed in location from time to time and covered by snipers.

2.--The G. H. Q. 1. line.

3.--The G. H. Q. 2. line.

4.--The Intermediate line.

5.--The Brown line.

6.--A series of centers of resistance at the following points (at present in process of construction):

(*a*) Elzenwalle--H. 36. c. 57. 94.--Foundation laid for 2 Moir M. G. emplacements, trenches, dug, but not revetted.--No wire.

(*b*) Gun Pits.--Wickham Market.--H. 29. d. 57. 57. (with rifle pit at H. 29. d. 60. 70. not yet commenced).--Complete except 16,000 yds. wire.

(*c*) Hemelryx.--Old Hickory Pill Box.--H. 29. b. 20. 05.--40@ complete.

(*d*) Cafe Belge Tunnel.--H. 30. a. 12. 25.--Deep Dugout.--Dugout sap and O. P. completed.--Trenches to be dug, wired and revetted.

(*e*) Segard Wood.--H. 30. d. 22. 48, with rifle posts at H. 30. d. 29. 59. and H. 30. d. 31. 62.--Complete except wiring of L. G. posts and strong point.

(*f*) Junction of G. H. Q. 2 line and Bellegord Avenue.--H. 30. b. 85. 85.--No work done.--Incomplete.--Small point occupied by Infantry and Lewis Gun.--No wire.

(*g*) Segard Water Tower.--H. 30. a. 96. 46.--Rifle posts and L. G. posts in G. H. Q. 2 at H. 30. a. 80. 41., and H. 30. a. 86. 46. Complete except wiring.

6. (*1*) During Normal Period:

Regimental Headquarters at Assam Farm.--H. 22. a. 2. 6.

The right Battalion, with Headquarters at H. 30. a. 80. 85., disposed two (2) Companies in the front line, distributed in depth.

The right front Company, with Headquarters at H. 36. c. 95. 85, has two (2) Platoons in the front line, and G. H. Q. 1 line; three (3) squads in the Gun pits (H. 29. d. 57. 57.), two (2) Platoons, less three (3) squads in the Ravine (H. 23. d. S. W.).

The left front Company, with Headquarters at H. 30. d. 9. 85. 85., has two (2) Platoons in the front line and G. H. Q. 1 line; one (1) Platoon in the Domino trench (H. 30. a.) and about White House (H. 24. c. S. W.) one (1) Platoon in the Ravine. H. 23. d. S. W.

The two rear Companies, with Headquarters at H. 22. d. 9. 0., and Smythe Farm (H. 21. d. 8. X. 3.) in posts forming a defensive flank facing S. and S. E., between Opium Farm (H. 21. s. 9. 7.) and the Vijverhoek (H. 29. a. 35. 30.).

The left Battalion, with Headquarters at H. 24. c. 6. 2., is disposed:

One (1) Company, with Headquarters at H. 30. d. 2. 5. in the front line and G. H. Q. 1 line.

One (1) Company, with Hqrs., at H. 30. a. 95. 55., has two (2) Platoons in G. H. Q. 2 lines and Intermediate line in H. 30. b. and I. 25. a. and two (2) Platoons in the Ravine in H. 23. d. and b. and H. 24. a. S. W.

Two (2) Companies in the Brown line H. 22. a. and c. H. 16. c. with Company Headquarters at Bolton Farm, H. 22. c. 4. 9. and H. 16. c. 3. 2.

The Support Battalion, with Hqrs. at G. 11. d. 2. 2. has:

One Company, Hqrs. at Pigment Farm G. 13. d. 0. 5. in the Viamertinghe Switch Line in G. 13. d.

Two Companies, with Hqrs. at G. 18. c. 25. 50. and G. 18. a. 3. 2. In the Yellow Line--Branch Line in G. 18.

One Company, with Hqrs. at G. 11. d. 3. 5. and G. 11. d. and G. 17. b.

The L. T. M. Battery, with Hqrs. at H. 23. b. 9. 1. has four forward guns and two (2) reserve guns.

The One Pounder Platoon has Hqrs. at H. 23. b. 9. 1.

The Pioneer Platoon has Hqrs. at H. 23. b. 9. 1.

2. In the event of definite warning of a hostile offensive on a large scale a message "Assume The Alert Position" will be sent out from Regimental Headquarters.

Upon receipt of this message the following movements will be carried out as soon as practicable.

(a) The Centers of Resistance will be manned by their full complements as follows:

Bn.	*Centre of Resistance*	*Garrison*	*Furnished by*
Right	(a)	1 Platoon	Right Company
Right	(b)	1 Platoon less 1 Rifle Section	Right Company
Right	(c)	2 L. G's. and Teams 1 Squad Riflemen	Right Company
Right	(d)	1 Platoon and 1 Rifle Section not needed in (b). Remainder of Platoon not needed in (c).	Right Company
Left	(e)	1 Platoon	Front Company

Left	(f)	1 Platoon	Front Company
Left	(g)	1 Platoon	Front Company

(*b*) All troops forward of the Intermediate line, with the exception of garrisons of Centres of Resistance designated in sub-paragraph (2) (a), will be withdrawn to the Intermediate line where they will take up positions in depth, the Right Battalion holding between the Regimental Boundary and Den Groenen Yager Cafe Nelge Road, inclusive, the Left Battalion between Den-Groenen Yager-Cafe Belge Road, exclusive and the Regimental Northern Boundary.

One Company of the Right Battalion will be distributed in and about the Trench line from H. 29. a. 40. 05. to H. 23. d. 05. 20.

One Company of the Right Battalion will dig in about the line H. 29 a. 0. 5. to H. 23. c. 6. 6.

The exact positions will be selected by the Battalion Commanders concerned and tapped forthwith, but no digging will take place unless the message referred to above is sent out.

One Company of the Left Battalion, plus the one Platoon of the Front Company not required to garrison Centres of Resistance, will be distributed in and about the Trench line from H. 23. d. 0. 5. to H. 24 a. 2. 2.

One Company of the Left Battalion will dig in about the line H. 23 c. 6. 6. to. H. 23. b. 0. 4.

The exact positions will be selected by the Battalion Commanders concerned and tapped forthwith, but no digging will take place unless the message referred to above is sent out.

The Right Rear Company of the Right Battalion, and the Left Rear Company of the Left Battalion will constitute the Regimental Reserve.

The Rear Battalion will, with one (1) Battalion of the 120th Infantry Regiment, constitute the Brigade Reserve.

Right Battalion Headquarters will move to:

Left Battalion Headquarters will move to:

7. POLICY OF DEFENSE:

(*1*) Defense of Sector will be active. "No Man's Land" will be patrolled constantly with the object of:

(*a*) Maintaining superiority over enemy patrols and thereby insuring control of "No Man's Land."

(*b*) Locating hostile forward posts.

(*c*) Insuring a constant flow of prisoners. This is all important, as prisoners and the consequent identification of hostile units form the surest source of information with regard to the enemy's intentions.

(2) DURING NORMAL PERIODS:

Any incursion of the enemy into our lines will be met by immediate counter-attack by Platoons, or Companies, held in readiness for this purpose. With this object, each Company earmark a Platoon, each Battalion a Company.

These counter-attack units will:

(*a*) Be in possession of written instruction as to their action in case of attacks.

(*b*) Carry out a thorough reconnaisance of the area held by the unit or formation to which they belong, and the areas on the immediate flanks of that unit or formation. On the reconnaisance, and a thorough study of the ground to be operated over, the success of the immediate counter-attack will mainly depend.

(3) IN THE EVENT OF A HOSTILE OFFENSIVE ON A LARGE SCALE:

All the ground will be held. There will be no retirement and every position occupied will be fought and held to the last man and last round. There will be no withdrawal except by order to withdraw from the authority who assigned the position to be held.

It is to be clearly understood that the object of the move referred to in paragraph 6-(2) will be:

(*a*) To lessen the number of casualties which may be expected from the enemy's preliminary bombardment of our forward areas.

(*b*) To break up his attack and inflict the heaviest losses upon his assaulting troops by the Garrison of Centres of Resistance (which in every case will be held to the last.)

(*c*) To keep as large a proportion as possible of troops in hand to counter-attack and recover the original line at the earliest moment.

That the move referred to in paragraph 6-(2) is only to be carried out after receipt of orders from Regimental Headquarters to do so.

(4) IN THE EVENT OF ANY CENTRE OF RESISTANCE BEING LOST:

The C. O. of the Battalion concerned will order an immediate counter-attack to be made to recover it.

Schemes for counter-attacking these Centres of Resistance will be prepared beforehand; in no case will the garrison of any Centre of Resistance be used for the purpose of carrying out a counter-attack to recover another one.

The security of the Immediate line will be of primary importance and must be provided for at all times in considering the force to be detailed for counter-attack forward of that line.

Nucleus garrisons of not less than three (3) Platoons each Battalion section will therefore be told off to selected portions of the Intermediate line and these will, on no account, be used to counter-attack forward of this line until relieved by troops sent up from behind.

8. ACTION OF TWO (2) REGIMENTAL RESERVE COMPANIES:

These two (2) Companies will be available:

(*a*) To deliver a counter-attack against any portion of the Regimental Front captured by the enemy.

(*b*) To hold a defensive flank facing S. W. along the line Opium Farm (H. 21. c. 9. 7.) Vijverhoek (H. 29. a. 2. 3.).

(*c*) To hold the Brown line within the Regimental area. All officers of the two (2) Reserve Companies will make themselves thoroughly acquainted with the whole of the Regimental area and areas on the immediate flanks.

In the event of these Companies being ordered to move forward, four (4) Lewis Guns per Company will be left behind as a nucleus garrison of the Brown line and Opium Farm-Vijverhoek Line, unless troops have previously arrived to take over the defense of this line.

The Lewis Gun positions will be carefully selected by the Company Commanders concerned and marked by notice boards as "Primary Lewis Gun Positions".

The positions for the remaining Lewis Guns of the two (2) Companies will also be selected and marked by notice boards.

9. The Support Battalion will be available for any operations in the Brigade area. Officers of this Battalion should thoroughly familiarize themselves with the area by personal reconnaisance.

10. POLICY OF WORK:

The Regiment is responsible for work on all defenses in the sector forward of the Intermediate line. The 105th Engineers will assist in this work.

For the present the working program will be as follows:

(*a*) Preparation and construction of Centres of Resistance.

(*b*) Improvement of trenches actually occupied by troops and communication trenches. Working parties forward of the Intermediate line will, in event of sudden hostile attack, immediately report to the nearest Infantry Commander and place themselves at his disposal. There will be no movements towards the rear.

11. All officers and N. C. O.'s will familiarize themselves with the location of Company, Battalion and Regimental ammunition dumps. The map co-ordinates of the same will be furnished on separate annes. The ammunition supply will be maintained at the minimum requirements at all times.

12. Scheme of Defense for T. M. Battery will be arranged by Battery Commander and reported to these Headquarters.

13. Regimental Headquarters will remain at Assam Farm.--H. 22. a. 3. 6.

14. Acknowledge.

J. VAN B. METTS,

Colonel Commanding 119th Infantry.

Distribution:

1--30th Division.

2--60th Brigade.

3--120th Infantry.

4--105th Infantry.

5--105th Engineers.

6--G-1, 30th Division.

7--119th Infantry

8--1st Battalion.

9--2nd Battalion.

10--3rd Battalion.

11--T. M. Battery.

12--37 M. M. Platoon.

13--Headquarters Co.

14--M. G. Company.

15--Signal Officers.

16--Transport Officer.

17--Supply Officer.

18--Surgeon.

19--File.

20--Diary

A TRUE COPY:

Capt. 119th Infantry.

On the night of August 26-27, 1918, the 2nd Battalion, 119th Infantry Regiment was relieved by the 1st Battalion, 119th Infantry Regiment, in the left Regimental front. Upon the completion of the relief, the 2nd Battalion withdrew to reserve position vacated by the 1st Battalion. Disposition of the 1st Battalion was as follows:

Company "C" in G. H. Q. 1 trench protecting Battalion right front. Two platoons Company "D" on left of Company "C" protecting Battalion left front. Two other platoons of Company "D" on left of Segard Woods. Company "B" in the Ravine. Company "A" in the Brown Line near Assam Farm.

About the night of 27-28 of August, Company "M" was relieved by Companies "I" and "L", and "M" Company withdrew to Ottawa Farm. Company "I" was occupying the right front of 3rd Battalion with two platoons in G. H. Q. 1 line and the remainder of Company on Anzac Ridge. Company "L" occupied the left front of 3rd Battalion with two platoons in G. H. Q. 1 line, one platoon in Domino Trench and the other platoon in the Ravine. Company "K" was in the Brown Line near Smythe Farm.

Early on the morning of August 31st the Commanding Officer, 119th Infantry received telephonic communication from the Division Commander that the Germans had evacuated, or partly evacuated Mount Kemmel, and that the British had advanced to occupy the evacuated section.

A general retirement of the Germans was now anticipated on this front due to their heavy losses and the rapid advance of the Allies further south. That morning about 9:00 a. m., the Division Commander instructed the Commanding Officer 119th Infantry to push out strong combat patrols over his entire front and gain contact with the enemy. Patrols were sent out about 10:30 a. m., and advanced about 100 yards when stout opposition was met as the enemy machine guns came into operation making it impossible for these patrols to advance much further, until additional aid was received from supporting platoons, establishing a general firing line. After the line had been strengthened the advance began again that afternoon about 6:00 p. m., and continued at intervals throughout the night and the following morning.

During the fighting that day two of our men were captured and five or six killed. Several of the enemy were killed and one captured. From wounds seen on the bodies of the dead Germans there appeared evidence of bayonet fighting and hand-to-hand contact with the enemy.

At 7:30 a. m., on the morning of August 31st, the troops composing the attack of the 119th Infantry Regiment had reached the line Voormezeele-Switch in I. 31. 6. and I. 32. A. Lock 8, inclusive, and thence along existing front line, capturing the town of Voormezeele. Upon reaching this point orders were received to consolidate and hold the position gained. Both flanks of the Regiment were exposed and Liaison could not be obtained with the units on the right and left. The 108th Infantry Regiment of the 27th American Division being on the right and the 120th Infantry Regiment on the left. Upon the consolidation of this position one platoon of Company "I" was sent up to protect the right flank of the 3rd Battalion, Companies "K" and "M" moved up to G. H. Q. 1 line. On the night of September 1-2, Companies "E" and "G" of the 2nd Battalion relieved Companies "L" and "I" of the 3rd Battalion in the advance position. Companies "F" and "H" remaining back in reserve.

Company "C" of the 1st Battalion was moved back to the Ravine. Two platoons of Company "B" in close support were moved up and relieved the two platoons of Company "D" on the left half of 1st Battalion in the advance position. Company "B" was now holding the entire front of 1st Battalion. Two platoons of Company "D" withdrew to G. H. Q. 1 line and the other two occupying small trenches between there and the new line.

Two platoons of Machine Gun Company, 119th Infantry Regiment came back in the line August 31st for the attack; one platoon was with the 1st Battalion and one with the 3rd Battalion. The other platoons remained in reserve. (It is well to note here that the 119th Infantry was the first American unit to take over a front line sector in Belgium, the first to make an advance, the first to capture a prisoner and the first to capture a town.)

On August 29th, 1918, orders were issued to the effect that the 118th Infantry Regiment (American) would relieve the 119th Infantry Regiment (American) on night of 31st, August and 1st September. Advance parties were sent to join the 119th Infantry in order that they might become acquainted and familiar with the front which their regiment was ordered to take over in a couple of days. This relief never took place on account of the advance made by the 119th Infantry Regiment and the advance party of the 118th Infantry Regiment rejoined their organization September 1st, 1918.

During the training, occupation and operation in the Canal sector the following number of officers and men were either killed, wounded or captured.

	Killed	*Wounded*	*Died of Wounds*	*Missing*
Officers	3	8		
Men	64	208	12	2

On September 2nd, 1918, the following orders were issued relieving the 119th Infantry Regiment from the line:

HEADQUARTERS, 119TH INFANTRY
American Expeditionary Forces
SEPTEMBER 2, 1918

SECRET

FIELD ORDER No. 22

COPY No. 00

Reference Map:

--Sheet 28, N. W. 4

--Sheet 28, S. W. 2

1. On the night 3-4 September, 1918, the 119th Infantry Regiment, will be relieved by the 106th British Brigade, in the Right Sector of the 30th Division (Canal) Sector, in accordance with attached table.

2. The Right Front Battalion will be relieved by the 19th H. L. I. The Left Front Battalion will be relieved by the 12th H. L. I. Companies L and I, by the other units occupying the positions occupied by the Support Battalion of the British unit relieved by this Regiment will be relieved by the 17th Royal Scots.

The Companies of the Battalion in Reserve to be announced later.

The Headquarters Company will be relieved by corresponding units of the 106th British Brigade.

The Machine Gun Company under arrangements to be made by Divisional Machine Gun Officer.

All details of relief will be made by the Battalion and Separate Company Commanders concerned.

Command will pass to C. O.'s of relieving units upon completion of Relief.

3. Guides will be furnished by Battalions occupying the Right and Left Sub-Sectors at the following rate:

1--Per Battalion Headquarters.

1--per Company Headquarters.

1--per Platoon.

Other guides will be furnished as follows:

Companies "I" and "L"	1 per Company Headquarters. 1 per Platoon.
Other units	1 per T. M. Battery.

All guides will assemble at Road Junction H. 22. a. 35. 85., at 8:30 p. m., September 3rd, 1918.

4. The Right Front Battalion and Left Front Battalion will send each the following detail to Dirty Bucket Camp, Sheet 28-A. 30., to arrive not later than 5:00 p. m., to arrange for distribution troops of this command therein.

5. Upon completion of Reliefs, Organization-Commanders will turn over maps defense schemes, information of the enemy, program of work, trench stores, ammunition dumps, grenades, etc., and take property receipts thereof.

6. Completion of Reliefs will be reported to these Headquarters using Code Word "Granted".

7. Regimental Headquarters will close at Assan Farm upon completion of Relief and will open at Dirty Bucket Camp, 28-A. 30.

J. VAN B. METTS, *Colonel Commanding.*

Distribution:

Copy No.

1--30th Division.

2--60th Brigade.

3--120th Infantry.

4--105th Infantry.

5--1st Battalion.

6--2nd Battalion.

7--3rd Battalion.

8--M. G. Co.

9--Surgeon.

10--Intelligence Officer.

12--Signal Officer.

13--Supply Officer.

14--105th Engineers.

15--War Diary.

15--File.

16--106th British Brigade.

The following morning the Regiment moved by light cavalry from Dirty Bucket Camp to Road Camp at St. Jan-tor Biegen (Belgium) and remained there one day and a half, resting and bathing.

September 5th, 1918, the following orders were received:

HEADQUARTERS, 119TH INFANTRY
American Expeditionary Forces

SECRET

FIELD ORDER No. 23

Reference Map:--Sheet 27, 19 S. E.

1. This Division is being transferred to the Third (British) Army. The move will be made by rail.

2. (*a*) This Regiment with attached troops, will entrain on the night of 6-7 September in accordance with entraining table attached hereto

(*b*) Entraining Station Heidebeek (X 27. a. 2. 4.), duration of journey, about 12 hours.

(*c*) Supervision of loading at railway station will be directly under the supervision of Major C. H. Mahon, 118th Infantry Regiment. No restrictions on roads for movement of troops to entraining station.

(*d*) Allotment of billets in detail and orders for detrainment will be given the Commanding Officer of each train on arrival in new area.

(*e*) Billeting parties of 1 officer and 2 men from each Battalion have proceeded to the new area.

(*f*) Officers in command of units arriving at the railhead will have prepared prior thereto a detail strength report in duplicate, giving the number of officers and men separately; the number of vehicles and kinds, the number of L. D., and H. D. animals separately and number of pieces of extra baggage and equipment, if any authorized. One copy of this report will be handed to R. T. O. and the other to the officer in command of entraining troops at their respective stations.

3. Information concerning supplies, refilling, changes of railhead and evacuation of sick and wounded will be furnished units upon arrival at detraining stations.

4. Battalion and separate Company Commanders will be held responsible for the police of camp and billets upon departure from this area, and that no property belonging to organization or individual is left behind.

5. Regimental Headquarters will close at Road Camp at 5:00 p. m., September 6th. Opening will be announced later.

J. VAN B. METTS, *Colonel, 119th Infantry*

Copies to:

1--30th Division.

2--60th Brigade.

3--1st Battalion.

4--2nd Battalion.

5--3rd Battalion.

6--Headquarters Company.

7--M. G. Co.

8--Supply Company.

9--Surgeon.

10--Intelligence Officer.

11--Signal Officer.

12--War Diary

13--File.

14--Capt. Wright.

Train No.	*Departs*	*Entrain at Heidebeek*
6	Sept. 6th, 7:30 p. m.	119th Infantry Regiment Headquarters. Headquarters Company. Supply Company. Machine Gun Company. Medical and Ordnance Detachment. Field Signal Outpost Section.
7	Sept. 6th, 10:50 p. m.	1st Battalion, 119th Infantry. Less Co. "D" Rolling Kitchen and team.
8	Sept. 7th, 1:50 a. m.	2nd Battalion, 119th Infantry. Less Co. "H" Rolling Kitchen and team.
9	Sept. 7th, 4:50 a. m.	3rd Battalion, 119th Infantry. Less Co. "M", Rolling Kitchen and team.
10	Sept. 7th, 7:50 a. m.	Co.'s "D", "H" and "M", 119th Infantry. With Rolling Kitchens and teams. 134th Field Ambulance.

Upon the transfer from the 2nd British Army, the Regiment proceeded by rail to the St. Pol Area, arriving there the morning of September 7th, 1918. Regimental Headquarters and Headquarters Company were located at Maisnel St. Pol. 1st Battalion and Companies "C" and "D" at Neuville au Cornet, Companies "A" and "B" at Monts En Tornois. 2nd Battalion at Buneville. 3rd Battalion at Moncheaux. Machine Gun Company at Tachincourt. Supply Company at Herlin Le Sac.

This Area was located near St. Pol, France, a town of probably 30,000 inhabitants and was known as a rest and training sector. Here, for the first time in two months, the Regiment was out of the Enemy Artillery Range. The men enjoyed the experience. The time there was devoted to recreation, however, training for future work did not suffer. One of the points greatly stressed was the Infantry attack assisted by tanks. An opportunity was given to three companies to engage in assimilated attack with the Tanks while the other companies of the Regiment observed their tactical use. This training though not appreciated at the time, proved of inestimable value in the subsequent operations of the Regiment.

On the 17th day of September, 1918, the Regiment moved by rail pursuant to the following order to the Puchevillere Area:

HEADQUARTERS, 119TH INFANTRY
American Expeditionary Forces
FRANCE, SEPTEMBER 16, 1918

SECRET

FIELD ORDER No. 24

Reference Map:--Lens 1-100.000

1. This Division will move to the Puchevillers area on the 17th and 18th instant. Troops will move by railway. Transport will move by road on the 17th instant, staying on the night of the 17th at Bougeumaison. This Regiment will move on the 17th instant.

2. (*a*) Entraining will take place in accordance with entraining table attached hereto:

(*b*) Entraining Stations will be as follows:

Petit Houvin	3D 86. 90
Bryas	2E 60. 91

(*c*) Detraining Stations as follows:

Belle Englise	6F 77. 88
Rosel	6E 03. 97

Duration of journey about 6 hours.

(*d*) Route of march will be the shortest route from present location to entraining point.

(*e*) Allotment of billets in detail and orders for detraining will be given the Commanding Officer of each train upon arrival in the new area.

3. Information concerning supplies, refilling, change of railheads and evacuation of sick and wounded will be furnished units upon arrival at detraining station.

4. Battalion and Separate Unit Commanders will be held responsible for police of camps and billets upon departure from area and that no property belonging to organization be left behind.

5. Regimental Headquarters will close at Maisnil St. Pol at 5:30 a. m., Sept. 17th, 1918 Opening will be announced later.

J. VAN B. METTS, *Colonel, 119th Infantry.*

Copies to:

1--C. G., 60th Brigade.

2--C. G. 30th Division.

3--C. O. 1st Battalion.

4--C. O. 2nd Battalion.

5--C. O. 3rd Battalion.

6--C. O. Hq. Co.

7--C. O. M. G. Co.

8--C. O. Supply Co.

9--Signal Officer.

10--Surgeon.

11--Intelligence Officer.

12--War Diary.

13--File.

ENTRAINING TABLE, To Accompany Field Order No. 24. Sept. 16th, 1918.

Date	*Hour*	*Entrain at Petit Houvin*	*Route*
Sept. 17	8:00 a. m.	Headquarters, 119th Infantry	No Instructions
Sept. 17	8:00 a. m.	Supply Company	No Instructions
Sept. 17	8:00 a. m.	Machine Gun Company	No Instructions
Sept. 17	8:00 a. m.	Headquarters Company	No Instructions
Sept. 17	8:00 a. m.	Field Signal, G. S. 119th Infantry	No Instructions
Sept. 17	8:00 a. m.	2nd & 3rd Bns., 119th Infantry	No Instructions
		Entrain at Bryas	
Sept. 17	8:00 a. m.	1st Battalion, 119th Infantry	No Instructions

All trains from Petit Houvin detrain at Belle Englise.

All trains from Bryas detrain at Rosel.

All troops will arrive at entraining point 30 minutes before scheduled departure of train.

Upon the arrival at Puchevillere Area the Regiment was located as follows:

Regimental Headquarters and Headquarters Company at Lealvillers; 1st Battalion at Clairfage; 2nd Battalion at Varennes; 3rd Battalion at Harponville; Machine Gun Company at Dragoon Camp; Supply Company at Toutencourt.

Here a most intensive training schedule was carried out over selected trenches and terrain preparatory to the attack on the Hindenburg Line, the importance of which was not realized until about ten days later.

September 22nd, the following Regimental Order was issued:

HEADQUARTERS, 119TH INFANTRY
American Expeditionary Forces
FRANCE, SEPTEMBER 22, 1918

SECRET

FIELD ORDER NO. 25

Reference Maps:

--Lens, 1-1000.000

--Amiens, 1-100.000

--St. Quentin, 1-100.000

1. The 30th Division is being transferred to the Fourth British Army.

2. This Regiment will move by bus on the night of 23-24 September, in accordance with attached movement table. All horse transport is moving by road on the night of 21-22 September, staging night of 22-23 September.

3. Billeting parties under command of Division Billeting Officer proceeded to new area at 9:00 a. m., 21st of September.

4. Guides will meet debussing organizations and conduct them to their camps or billets. Administrative details will be announced later.

5. Units leaving nights 23-24 of September will carry Iron Rations only.

6. Regimental Headquarters will close at Lealvillers at 5:30 p. m., 23rd of September, and open at time and place to be announced later.

J. VAN B. METTS, *Colonel, 119th Infantry.*

Copies to:

1--C. G. 30th Division.	9--Headquarters Company.
2--C. G. 60th Brigade.	10--Surgeon.
3--C. O. 120th Infantry.	11--Intelligence Officer.
4--C. O. 105th Infantry.	12--Signal Officer.
5--C. O. 1st Battalion.	13--Supply Officer.

6--C. O. 2nd Battalion.	14--105th Engineers.
7--C. O. 3rd Battalion.	15--War Diary.
8--M. G. Company.	16--File.

MOVEMENT TABLE.

Route.--Headauville-Bouzincourt-Albert S. edge of Fricourt-Carnoy Road to Cleary thence to Tincourt Area.

Remarks.--Head of column at trail intersector 300 yds. N. of Headauville. Clear Headauville at 9:30 p. m. Loading completed by 9:00 p. m. Two Motor Ambulances will accompany the column. Debussing point to be designated by the Brigade Commander.

This movement was carried out according to instructions and the Regiment arrived in the Tincourt Area about 6:00 a. m., September 24th, and after a four mile hike was camped in the southern edge of a wood immediately N. W. of Tincourt.

Here the Regiment had reached shell fire again as one could be heard whizzing overhead several times each day and everyone was beginning to realize that they were drawing close to the fighting line every hour on account of the rapidity and secrecy of order and movements.

That afternoon the Regiment moved up a little closer to the line and was distributed in sunken roads, chalk cliffs, etc., throughout the vicinity of Roisel.

Here several Australian Officers and N. C. O.'s were attached to the Regiment, to assist with their advice and aid in any way possible for the coming operations, as they were thoroughly familiar with this sector of the line.

The Regiment remained at Roisel three days, making their final plans for the attack on September 29th, against the Hindenburg Line.

September 26th, the following Regimental Order was issued:

HEADQUARTERS, 119TH INFANTRY
American Expeditionary Forces
FRANCE, SEPTEMBER 26, 1918

SECRET

FIELD ORDER NO. 27

COPY NO. 00

Reference Maps:

--Sheet 62-c. N. E. 1-20.000

--Sheet 62-b. N. W. 1-20.000

1. The 60th Infantry Brigade (less 115th M. G. Bn. and Regimental M. G. Co.), will relieve the 118th Infantry in the present sector on the night of 27th-28th of September.

2 (*a*) The 119th Infantry will relieve the troops of the 118th Infantry holding the area bounded on the North by a certain line running through A. 26, C. 9,9 and on the South by a line running through L. 16. B. 0. 9. G. 8. 6. 6. 6., to the road along Sentinel Ridge towards Bellicourt, in accordance with attached table "A"--Relief to be completed by 2:00 a. m.

(*b*) Inter-Battalion Boundary: Northern point of Bobby Farm (L. 11 a. 9. 5.), inclusive (to Right Battalion)--G. 2. c. 0. 4.

(*c*) The 1st Battalion will take over the right sub-sector. The 2nd Battalion will take over the left sub-sector. The 3rd Battalion will take up supporting position, to be selected by Battalion Commander. The Headquarters Company will relieve portions of the corresponding unit of the 118th Infantry. The M. G. Co., relieved the M. G. Co., 118th Infantry on the night of 26-27th of September under arrangements made by Division M. G. Officer.

(*d*) The 27th American Division will be on our left.

The 120th Infantry will be on our right.

3. (*a*) One man from the advanced party of each Battalion will return to his unit for the purpose of conducting them to the point where guides will meet them.

(*b*) All defense schemes, maps, trench stores, etc., will be taken over by units concerned and list of same with location forwarded to these Headquarters the morning after the relief.

(*c*) Active patrolling of "No Man's Land" must be the policy of the Regiment. Control of this territory will be secured and kept.

(*d*) All officers and men must be warned that in case of capture, no information other than name and rank will be given.

(*e*) Attention is directed to the fact that owing to our own bombardment enemy retaliation may be expected. Advantage must be taken of all available cover and any congestion must be prevented in order to avoid needless casualties.

4. (*a*) Command of Left Sub-Sector will pass to Commanding Officer, 119th Infantry, upon completion of relief.

(*b*) Completion of relief by each Battalion will be reported to these Headquarters by the Code Word "Fritz", telephone being used, if available.

(*c*) All details for relief not specifically provided for herein will be arranged between Battalion Commanders of the units concerned.

(*d*) Regimental Headquarters will close at its present location at 8:00 p. m., and open at L. 5. d. 5. 9.

5. Acknowledge.

J. VAN B. METTS,

Colonel, 119th Infantry.

Serial	*Units*	*From*	*To*	*Relieving*	*Remarks*
1	2-Bn.	Present area	Line left subsector as defined in F.C.	118th Infantry in new sector	No Restrictions as to route. Head of column to pass line North and South through L. 16. Central at 8:00 p. m.
2	1-Bn.	Present area	Line right sub - sector as defined in F. C	118th Infantry in new sector	No Restrictions as to route. Head of column to pass line North and South through L. 16. central at 8:30 p. m.
3	3-Bn.	Present area	Line support position	118th Infantry in new sector	Route head of column to pass line North and South through L. 16. central at 9:00 p. m.
4	Hqrs. Co	Present area	Line	118th Infantry in new sector	

The relief to be completed by 2:00 a. m.

All forward movement will be made in small parties and at full distance. Troops will not march in column of squads.

Troops will keep entirely to areas of units they relieve.

The usual anti-draft precautions will be taken.

Troops will march off roads as much as possible.

After taking over the line on the night of the 27-28 of September until the morning of the attack, great pains were taken to instruct each individual soldier in his particular duty.

The supporting Artillery which consisted of nine Field Artillery Brigades and five Heavy Artillery Brigades, kept up a continual fire harrassing the enemy and destroying his defense. The Artillery preparation covered a period of 48 hours prior to the attack.

On the night of September 28-29th, the tape was laid by the 105th Engineers. Due to casualties, however, this task was completed by officers of this Regiment. The Tanks which were to assist in the attack were led into position; and the Infantry took their place upon the tape in the following order: 1st Battalion, commanded by Capt. Otho Robinson, with

"A" and "D" assaulting Companies and Companies "B" and "C" in close support. 2nd Battalion, commanded by Capt. Graham K. Hobbs, with "F" and "G" assaulting Companies with "E" and "H" in close support. 3rd Battalion, commanded by Major William S. Privott, with the exception of "M" Company, which was broken up into carrying parties, was assigned as "moppers up", to follow the two front line Battalions and clean up all the trenches, dug-outs, and machine gun posts, taken by them. "I" and "L" Companies were to mop up the Tunnel alone. Company "K" assisted in the "mopping up", but was later called upon to protect the left flank of the 2nd Battalion.

The following order was issued governing the attack:

HEADQUARTERS, 119TH INFANTRY
American Expeditionary Forces
FRANCE, SEPTEMBER 27, 1918

SECRET

FIELD ORDER No. 28

Reference Maps:

--Sheet 62-b, N. W. 1-20.000

--Sheet 62-c, N. E. 1-20.000

1 (*a*) On a date and hour to be notified later, the Australian Corps with The American II Corps affiliated will attack the enemy's position in the Hindenburg line, in the sector near Nauroygouy.

The 27th Division (American) will attack on the left of the 119th Infantry.

The 120th Infantry will attack on the right of the 119th Infantry.

(*b*) Boundaries:

Northern Boundary: A. 26. c. 4. 7.--A. 24. a. 0. 2.

Southern Boundary: Line along road (incl.), on Sentinel Ridge to Bellicourt--thence along railway from G. 2. d. 0. 3. crossing Sunken Road at G. 5. d. 9. 7.--thence to objective (G. 6. c. 5. 8.).

Inter-Battalion Boundary: Goat line (incl.), to left Battalion to G. 3. b. 4. 4.--thence to Air Shaft in tunnel (G. 4. a. 5. 8.) (incl.), to left Battalion--thence to A. 28. d. 4. 1.--thence South to Cabaret Wood Farm (incl.), to left Battalion to A. 30. c. 2. 8.

(*c*) Objective: Line running through A. 23. b. 9. 1.--A. 24. c. 0. 0., A. 30. a. 2. 8.--thence South to G. 6. a. 4. 0. Under no condition will troops of this Regiment go beyond this objective.

2. (*a*) The 1st Battalion will attack on the right with the 2nd Battalion the left, each Battalion having two companies in the front line, and two (2) companies in close support, each company attacking in two waves of two lines each.

The 3rd Battalion will be in support and will mop-up the area captured between the starting point and the St. Quentin Canal Tunnel, inclusive, and will be detailed to mop-up the St. Quentin Canal and approaches thereto and the Sunken Road in A. 27 and G. 4.

One Platoon of the M. G. Co., will report to and be at the disposal of each Battalion Commander.

One 37 M. M. Gun will report to and be at the disposal of each Battalion Commander.

The Stokes Mortar will move in advance of the Support Battalion and assist in dealing with the situations that may arise and assist in the defense of the objective.

Dressing stations will be established under direction of the Regimental Surgeon who will report their locations to the units concerned.

(*b*) Supporting Troops:

1. Artillery:

There will be a preliminary bombardment by nine (9) F. A. Brigades and about five (5) Heavy Artillery Brigades, for a considerable period for the purpose of completing the enemy's demoralization and destroying his defenses. The Artillery will support the Infantry attack with a creeping barrage, including a considerable proportion of smoke shells (details of Artillery plan, time table, etc., will be issued later).

2. Tanks:

The Infantry will be proceeded by 12 tanks from the Fourth Tank Brigade. Battalion Commanders will furnish guides to conduct the Tanks from their forming up place to the Infantry Departure tape as already arranged. Battalion Commanders are responsible that these approaches be taped on the night of Day D-1 in sufficient time to permit the approach of the tanks.

3. Contact Infantry Aeroplanes will be available for communication from ground to Aeroplane. Flares and disk will be displayed only by front line troops when called for by these planes, when a Klaxon horn is sounded or a white, very Light is fired.

4. All Machine Guns of the Division (less Regimental M. G. Co.'s), will assist the advance by thickening the Artillery barrage.

(*c*) 1. The entire 60th Brigade will attack at H. Hour on D. Day. Distance between units to be gained while advancing. The right flank of the Brigade will be protected by an Artillery Smoke Screen.

2. Mopping up:

Mopping up will be done in accordance with map furnished the Battalion Commanders by Regimental Headquarters. Battalion Commanders will see that moppers up wear a distinctive mark, and that mopping up arrangements are worked out with the utmost care and detail. Special attention is called to the report on the St. Quentin Canal furnished the Battalion Commanders by Regimental Headquarters. Battalion Commanders will insure that definite bodies are allotted for duty at each entrance and exit to the tunnel, and that extra parties are detailed to search unknown entrances and exits. These troops must on no account be withdrawn from their task until their work is completely accomplished and will only be withdrawn upon the authority of an officer or Platoon Commander. A full supply of smoke bombs and grenades will be carried by moppers up.

3. (*a*) Formation for the attack.

Troops will assemble by 4:30 a. m., on the Tape line which is to be held by Engineers. Battalion Commander will insure that sufficient tape is laid running to the trenches to facilitate the forming of troops. The forming up of troops will be completed as quietly and quickly as possible and all troops will be massed on the tape as closely as possible in order to avoid the enemy barrage.

(*b*) Lifts of Artillery Barrage:

Barrage will lift at the following rates: 100 yards in three (3) minutes for the first 600 yards, thereafter 100 yards in four (4) minutes. On the barrage reaching 500 yards East of the St.Quentin Canal it becomes stationary for fifteen minutes, after which it resumes this advance at the rate of 100 yards every four (4) minutes.

(*c*) Tanks accompanying the Infantry will follow the barrage at approximately seventy (70) yards. The first wave will follow the tanks at approximately seventy (70) yards, action of tanks accompanying supporting troops will be dictated by the situation.

(*d*) Battalion Commanders will insure that specially selected men are employed on the flanks to maintain direction with the aid of prismatic compasses.

(*e*) The success signal will be a No. 23 Rifle Grenade bursting into three white lights.

4. (*a*) Care will be taken that signal is not fired until objective is reached.

(*b*) Under cover of the protective barrage, patrols will be sent forward and the positions prepared for defense.

5. Upon occupation of the line, immediate steps will be taken to establish liaison with flank units and maintain same throughout the operation.

6. Posts of Command before attack:

60th Infantry Brigade	L. 10. a. 1. 4.
119th Infantry	L. 5. d. 5. 9.
120th Infantry	L. 23. b. 7. 6.

7. Watches of Battalion Headquarters will be synchronized under Regimental arrangements and of Companies under Battalion arrangements.

8. Second phase of the attack:

The 5th Australian Division with its own Artillery and tanks will pass through the 60th Brigade after the 60th Brigade has attained its objective. After the whole Australian Division, including its reserves has passed through, the 60th Brigade will be re-assembled. Time and place will be notified later.

9. Communication Supply and Evacuation, as prescribed by the 30th Division orders regulating road movements and traffic control will be strictly observed.

10. General Remarks:

(*a*) When barrage halts on Canal for 15 minutes, every opportunity will be taken to re-organize units preparatory to resuming the advance.

(*b*) Ammunition, water, and tools can be obtained from Regimental Ammunition Dump at L. 10. a. 1. 7.

(*c*) All means and methods of communication will be employed to keep Regimental Headquarters informed of the situation, especially the location of the front line.

(*d*) There must be no doubt as final objective being reached.

J. VAN B. METTS,

Colonel, 119th Infantry.

Distribution of copies:

1--C. G., 30th Division.

2--D. G., 60th Brigade.

3--119th Infantry.

4--C. O., 120th Infantry.

5--1st Battalion.

6--2nd Battalion.

7--3rd Battalion.

8--C. O., Headquarters Co.

9--C. O., M. G. Company.

10--Regiment on Left.

11--1st Battalion, 4th Brigade Tank.

12--Signal Officer.

13--Intelligence Officer.

14--File.

15--War Diary.

HINDENBURG LINE.

To fully appreciate the importance of this operation it is here deemed necessary to explain somewhat in detail the magnitude of the undertaking. The Hindenburg Line so called because relied upon and established by the German General Von Hindenburg, was the last organized system of trenches to stop an Allied advance. So well was this system of trenches organized and so much faith pinned upon its impregnability, that a break through meant not only the destruction of the back-bone of the German defense system, but also the morale of the German Army. Another fact of extreme importance was that once the Allied Armies had penetrated the German line at this point, the enemy's army in Flanders would be completely bottled up with no means of escape. The attack was to take place on a Three Corps Front, with the Second American Corps, composed of the 27th and 30th Divisions in the Center, attacking opposite Bony and Bellicourt, the strongest point in the line.

"The Hindenburg Line at this point curves in front of the tunnel of St. Quentin. . . The line curving west of the tunnel consists of three main trench systems protected by vast fields of heavy barbed wire entanglements skillfully placed; this wire was very heavy and had been damaged very little by Artillery fire. The dominating ground enabled the enemy to bring devastating machine gun fire on all approaches. The lines had been strengthened with concrete machine gun emplacements. It contained at this point a large number of dugouts lined with mining timbers, with wooden steps leading down to a depth of about thirty feet with small rooms capable of holding from four to six men each. In many cases these dugouts were wired for electric lights. The large tunnel through which the Canal ran was of sufficient capacity to shelter a Division. This tunnel was electrically lighted and filled with barges on which the Boche lived connecting it with the Hindenburg Trench System and numerous tunnels. In one case a direct tunnel ran from the main tunnel to the basement of a large stone building which the enemy used for headquarters. Other tunnels ran from the main tunnel eastward to the city of Bellicourt and other places. This complete subterranean system, with its hidden exits and entrances, unknown to us formed a most complete and safe subterranean method of communication and re-enforcement for the German sector."

ATTACK ON HINDENBURG LINE.

The Regiment completed its formation on the tape by 4:30 a. m., Sunday, September 29th, 1918, and waited quietly the zero hour. At exactly 5:50 a. m., the most terrific barrage ever laid down on the Western Front opened and the Infantry following it as closely as possible, assisted by such a density of fog and smoke that a person could not be seen at a distance greater than five yards, succeeded in making its way through the strong fortifications assisting the 120th Infantry in the capture of the town of Bellicourt and the St. Quentin Canal, pushing forward towards its objective. The fog and smoke caused quite a bit of confusion. No one, except with the aid of a compass, could keep the proper direction, consequently the Companies became very much mixed and scattered, a situation which might have been more serious but for the work of the officers, the training of the men and the utter rout of the enemy.

Some units of the Regiment reached the final objective on the left but these were captured by the enemy as soon as they saw the gap in the line caused by the failure of the 27th Division to keep up with the advance. Because of this gap the Regimental left necessarily had to conform in order to prevent a retaking of the ground gained by a counter-attack. During the time that this left flank was exposed the Regiment suffered a loss by capture of three officers and several enlisted men and a great number killed and wounded.

A short time after the zero hour; prisoners were started to the rear and the road from Hargicourt to Bellicourt was practically filled with them during the entire day. These prisoners seemed overjoyed at the fact of their capture, especially by an American Unit. Quite a number of these were interviewed and readily gave any information desired.

Owing to the magnitude of the offensive, it was deemed necessary as a precautionary measure to further protect the left flank which had suffered so heavily during the engagement, against a possible counter-attack by the enemy. For this purpose one Battalion of the 118th Infantry was sent to the Regimental Commander to be used as he deemed necessary. He in turn sent the Battalion forward to occupy a trench which ran from the Canal in a westerly direction. This Battalion completed the occupation at about 12:30 a. m., the morning of September 30th, 1918. The counterattack never materialized.

Upon information gained from a German Battalion Commander interviewed at Regimental Headquarters, it was learned that when the enemy realized that the Americans had broken through their wire and first defenses, they withdrew the bulk of their reserves from the tunnel and reserve trenches, with the idea of making a stand at a position further to their rear. However, finding that the Division (27th American) on the left of the Regimental sector had failed to keep up with the advance and that the Regimental left flank was exposed, pushed additional troops in on the left. This move on the part of the enemy developed for the 119th Infantry Regiment, as it may be termed, a second battle for the day, on the then exposed left flank, between our advance line and the right flank of the 27th Division.

According to the scheme of attack, the Fifth Australian Division was to pass through the Regiment upon its reaching its final objective. Since the final objective was not reached, and since such a critical situation presented itself on the left flank, the Australians came into the line during the afternoon of September the 29th, and occupied the front line trenches along with the men of this Regiment, and it was not until the morning of the 30th, that they were able to make an advance. About the same time the Australians began their advance, orders were received to assemble the Regiment preparatory to withdrawal from the line for rest. It may be noticed that many of our men advanced with the Australians and had to be ordered to assemble in order that they might be relieved with the Regiment. The entire day of September 30th was consumed with a re-organization of the Regiment and the consolidation of the occupied position. Scouting parties were detailed to search every trench and dugout for the dead and wounded, numbers were found. Captured material was collected and placed along the road leading to Bellicourt. During the engagement, the Regiment suffered losses as follows:

	Officers	*Men*
Killed		146
Wounded	9	682
Died of Wounds	1	15
Prisoners	4	33
Missing		12

The field over which this fight took place, on the 30th day of September, presented a miserable appearance, as dead soldiers were scattered broadcast over its area. Shell holes were so numerous that one could not walk three steps without falling into one. Huge masses of barbed wire had been partly cut by the bombardment and the Tanks. Not a telephone pole nor a tree had been left standing by the sweeping Artillery fire. The town of Bellicourt was a complete wreck.

"WITHDRAWAL."

On the night of October 1st and 2nd, the Regiment withdrew to the location previously occupied by it in the Tincourt Area, and in compliance with orders on the morning of October 2nd, marched to the Herbecourt Area by the route of Tincourt-Monrouge-Peronne-Biaches, and went into camp shortly after noon the same day at a place, formerly the village of Biaches, now completely destroyed. This camp was on the Marne River and only a short distance west of Peronne.

The time here was occupied principally by resting, and a general policing of the personnel in the Regiment. The men enjoyed these two days of rest.

On October 5th, the Regiment marched back to the Tincourt Area and camped that night in Tincourt Wood. The following day orders were received to move up to the Bellicourt-Hargecourt Area and the night of the 6th and 7th of October, was spent camped in the trenches of the Hindenburg Line, which eight days previous were occupied by the Germans.

On October 7th, the march was continued to the town of Estrees where the Regiment halted for the night.

The night or the portion of night spent here was marked by a most terrific bombardment from the air. No casualties in the Regiment, however, resulted from this attack. Shortly after midnight verbal orders were received from the Commanding General, 60th Brigade, by the Commanding Officer of the Regiment to proceed to the vicinity of Premont and follow the attack of the 117th Infantry which would begin at 6:00 a. m., October 8th, passing through or "leap-frogging" the 117th at the railroad fill immediately west of Busigny.

"ATTACK OF OCTOBER 8th, 9th, AND 10th."

Leaving Estrees at approximately 3:30 a. m., the Regiment proceeded by marching to Premont and to its position, arriving there just as the barrage opened. Immediately it formed up in the following order: 2nd Battalion under command of Capt. Graham K. Hobbs, with "C" and "E" assaulting Companies and "G" and "H" the two supporting Companies occupied the left Regimental sector. The 3rd Battalion under command of Major William S. Privott and in the same formation with "K" and "M" Companies as the assaulting Companies and "I" and "L" Companies supporting, occupied the right Regimental sector. The 1st Battalion under command of Capt. Robert J. Lamb, was in support with Company "C" supporting 3rd Battalion and Company "D" supporting 2nd Battalion. Companies "A" and "B" were in reserve. The Regiment joined the 25th British Division on the left and the 120th Infantry Regiment on the right. Regimental Headquarters were established at Van le Prettre Farm. In this formation the line advanced approximately 1,500½ yards, on being held up here immediately west of the woods east of Premont, the 119th Infantry Regiment pushed through and continued the advance to its objective a short distance north-east of the town of Busigny, capturing the town and a large quantity of material, including a Battery of 5.9's and several field pieces, also liberating 1,800 French civilians who had been under control of the Germans for the four years of the war. Upon the entrance of the Americans into the village and before the Germans had been driven entirely out, the civilians were in the streets offering the soldiers hot coffee. They could not be too kind to their liberators. It seemed to surprise them greatly that the troops occupying their town then were friendly Americans. They thought that the British had freed them and it was sometime before they could believe the truth.

At about 4:30 the same afternoon the Regiment gained all of its objectives and immediately consolidated its position. Regimental Headquarters moved from Van le Prettre to Premont. This advancement was of 6 kilometers.

The attack order for the advance on the morning of October 9th, was received at Regimental Headquarters about 3:30 a. m., and transmitted to the Battalion Commanders and front line troops by 4:30 a. m. A runner was the only means of communication at this time. In the attack order the barrage was timed to open at 5:00 o'clock, but later was changed to begin at 5:30, and, had it not been for the fact that the attack orders were late in arriving at the front line, this difference of 30 minutes might have caused disastrous results.

The Regiment in the same formation as the previous day resumed the attack at 5:30 a. m., October 9th, with objective the Heights 500 yards east of the LaSalle River. This objective was selected for its tactical advantage, as it gave the enemy a dominating view of the country west of the LaSalle River, and from information gained seemed to be very strongly fortified.

The advance continued without meeting very strong resistance until the town of St. Souplet on the west bank of the LaSalle River was reached; about 3:00 o'clock that afternoon.

Here the enemy apparently had time to make some preparation and to bring up a few reserves, as the Heights immediately east of the River were well protected. Heavy Machine Gun and Artillery fire was received from the entire ridge covering the Regimental front. It was practically impossible to advance up this open incline without the sacrifice of a great number of lives. Repeated efforts were made to cross the River, and at a few points were successful but due to the nature of the terrain and the terrific Machine Gun fire, it was deemed necessary to dig in and hold the line along the west bank of the River, as a counter-attack was anticipated, but never materialized. The left flank of the Regiment at this time was connected with the 25th British Division but the right flank was exposed as the 120th Infantry had not kept up, causing the line to bend back in front of several fortified Machine Gun posts.

Confronting these conditions, the line was ordered to halt and make good a general line along the west bank of the LaSalle River and gain contact with the 120th Infantry, which was done later that night. During the day's operation the Regiment advanced to a distance of 4,000 yards and captured the towns of Escaufort, St. Souplet and St. Benin, liberating 706 French civilians. A large number of Machine Guns and several field pieces were captured with a large supply of ammunition.

That afternoon about 2:30 o'clock Regimental Headquarters moved from Premont to Busigny.

The night of the 9th-10th of October was spent in consolidating the captured position and to gain contact with the unit on the right which had been held up by Machine Gun fire. It was the plan that as soon as this connection had been made an Artillery Barrage was to be laid down on the whole front thus enabling the entire line to advance. The German position could not be taken without this preliminary preparation by the Artillery so strongly was it defended.

On the morning of October 10th, since the 120th Infantry had not come up on the line, the plan of the attack was not enforced but instead the Regiment merely exploited a short distance immediately forward of this position. After the hard fighting of the 8th and 9th, the men were completely worn out and ready for a rest. However, their high standard of morale had never waned. On the afternoon of the 10th, orders were received for the relief of the Regiment by the 108th Infantry Regiment of the 27th American Division. This relief, due to the fact that some of the units of the 108th Infantry became separated from the rest of the organization, was not completed until after daylight on the morning of the 11th of October. The troops moving from the front line positions could plainly be seen by the enemy who immediately fired on them with Artillery. The Regiment moved by marching back to Premont, where it remained until the same afternoon when it marched to the area around Van le Prettre Farm. Here it was billeted for the night of October 11th-12th. Due to a mistake, the allotment of billeting area by Corps Headquarters, the Regiment was compelled to move to the area surrounding Montbrohain, while Regimental Headquarters was established in the village.

In this area every advantage was taken to rest the men and to let them clean their clothing and equipment. A captured German bathhouse was utilized and every man in the Regiment was thus enabled to get a good hot shower bath. The Regiment remained here for five days.

"ATTACK OF OCTOBER 17th, 18th, AND 19th."

On the afternoon of the 16th of October, in accordance with instructions the Regiment moved to the woods immediately southwest of Busigny, where it bivouacked for the night of the 16-17th. Regimental Headquarters was established in the village. Before daybreak the morning of the 17th, in accordance with battle instructions, Headquarters 30th Division, the troops of the Regiment marched to its forming up area along a general line, eastern outskirts of La Haie Mennerese and bore on Vaux Andigny, arriving there at 5:00 a. m., zero hour. From this point, the Regiment was to follow the advance of the 59th Brigade under an Artillery Barrage, passing through it at zero plus six hours. The following disposition of troops was made: The 1st Battalion commanded by Capt. Robert J. Lamb on the left of the Regimental sector with "A" and "B" Companies assaulting, "C" Company in close support and "D" Company in support. 2nd Battalion commanded by Major Graham K. Hobbs on the right Regimental sector with "E" and "H" Companies assaulting, "H" Company in close support and "G" Company in support. The 3rd Battalion commanded by Major John Hall Manning, was in support. The Regiment was joined on the left flank by the 120th Infantry and on the right by the 1st British Division of the 9th British Corps. While following up the 59th Brigade, the Regiment passed through an extremely heavy counter barrage of the enemy.

At approximately 11:00 o'clock in the morning, the 119th Infantry Regiment took over the line from the 117th Infantry as it was held up by Machine Gun fire, and advanced to a railroad cut 2,000 yards west of the village of Ribeauville. Here the position was consolidated and the men dug in for the night. This position was very heavily shelled and quite a number of casualties resulted. The command of the sector officially passed from the Commanding General, 59th Brigade, to the Commanding General, 60th Brigade, at sundown the same evening. With the passing of the command the 59th Brigade supported the 60th Brigade in its advance.

Due to the fact that the British on the right and the 120th on the left had been unable to keep up with the advance and gain connection, the attack on the morning of the 18th inst., could not be made. During the entire morning harassing Machine Gun fire was distributed over the Regimental front. About 11:30 a. m., the British by aid of Artillery Barrage succeeded in advancing until connection had been established with the 119th Infantry. At the time connection was made with the British, the 119th Infantry attempted to advance on the town of Ribeauville, but this advance was met by such heavy Machine Gun fire that the right Battalion had to fall back to its original position. Practically no advance had been made by this Regiment up to 5:00 o'clock that afternoon, nor by the units on either flank.

About 7:30 o'clock information was received that the British on the right had moved up to the prolongation of our front line, and were continuing the advance. Information was also received that the 27th Division (American) had succeeded in pushing its line forward.

At 8:00 o'clock that evening orders were received from Commanding General, 60th Brigade, to the effect that the 119th Infantry would advance immediately, keeping liaison with units on right and left. The attack formation consisted principally of Combat Groups and Patrols joined by connecting files, which pushed out and continued the advance over very rough country, assisted only by the dim moonlight and compass. Owing to these conditions, very slow progress was made, as it was very hard to maintain connections.

Very little Machine Gun and Artillery fire was encountered until the line reached the eastern edge of Ribeauville, where shelling became heavier, and a great number of Mustard Gas shells were falling in the area.

The line continued to advance to the village of Ecaillon, which was reached about 1:30 a. m., October 19th. Orders were received to halt and secure connection with units on right and left, which was done. That night the town of Ribeauville was captured liberating one French civilian, and at Ecaillon five horses were captured in a stable with saddle equipment and harness.

About 3:30 that morning orders were received that the attack would resume at 7:00 o'clock on the entire Corps front, and to continue the advance as far as possible, no objective being given. The Regiment in practically the same formation as the previous day formed up and advanced in the general direction of Catillon, maintaining connection on its flanks. At 8:30 the town of Mazingheim was captured by 119th which continued to advance, until it had reached the high ground east of Mazingheim overlooking the Sombre Canal. Here very stout Machine Gun opposition was met, and the line held up for awhile. Supporting Machine Guns, Stokes Mortars, and One Pounders were sent up to protect the front from any counter-attack and did good work in putting out several enemy Machine Gun posts. Practically no advance was made beyond this high ground that afternoon, due to the very open country and the deadly fire from the German Machine Guns.

Orders were received to halt and make good the line held. The men and officers were now, after a period of three days' continuous fighting, very, very tired, having advanced a distance of practically five miles.

That night the Regiment was relieved by the _____ British and the Regiment retired to the town of St. Souplet where it was billeted the night of October 19th-20th. The following morning the Regiment moved by marching to Busigny, and October 21st marched to Joncourt. The march continued October 22nd, and that afternoon the Regiment was camped in small huts immediately west of Tincourt. October 23rd, the entire Regiment moved by rail from Tincourt to Heilly and marched from there to Contay Area near Amiens, where the units were billeted in the following towns: 1st Battalion at Contay; 2nd Battalion and 3rd Battalion at Behincourt; Headquarters Company at Baolincourt; Machine Gun Company at Contay; Supply Company at Hahoussoy; Regimental Headquarters at Esbart Chateau.

"REST AND TRAINING."

Upon the arrival of the Regiment in this area, every advantage was taken to give the men a thorough rest and also to equip them. The depleted ranks had to be filled by replacements, for the Regiment was in strength, a little larger than that of one Battalion. The Commanding Officer expressing the wish of all the officers and men of the Regiment, exerted every effort to have the old men of the Regiment who had been pronounced ready for duty from hospitals, returned to the organization. As a result of this effort quite a number of the old men were returned along with a few replacements to take the place of those killed and wounded. After the rest of a few days, a training schedule was inaugurated preparatory to fitting the organization for further fighting. Quite a number of maneuvers were included in the training schedule and the majority of them were carried out, assimilating open warfare.

Connected with the training a number of field meets were held, principally to relieve the men's minds from the battle front and to create a spirit of competition throughout the entire 2nd American Corps.

On or about November 17th, the 30th Division was relieved of further duty with the British Expeditionary Forces and transferred to the American Expeditionary Forces. With this transfer the Division was moved from its area near Amiens and sent further south with the Americans. On November 21st, 1918, this Regiment entrained at Corbie and moved by rail to the LaMans Area, detraining the following morning at Beaumont sur Sarthe. Regimental Headquarters, Headquarters Company, Machine Gun Company, Supply Company, 1st and 3rd Battalions, were billeted in the town, and the 2nd Battalion was billeted in the village of Vivian. This order was later changed, since the billeting space in Beaumont was entirely too crowded, and some of the organizations were moved into neighboring villages.

Training schedule was inaugurated here and the Regiment sustained its high standard by attaining a high degree of perfection in drill, ceremonies, training and equipment.

The citizens of Beaumont were most cordial to the Americans and exerted every effort towards their entertainment. The officers were extensively entertained in their homes. At a dinner given by the Mayor of the town, the Commanding Officer of the Regiment was presented with a "wreath of Laurel" in token of the splendid work of the Regiment during the fighting.

While in this area, a scheme of competition for the entire 2nd Corps was started. This scheme meant the selection of the best all round Regiment of each Division which were selected by means of competitive drills and inspection to determine the best Regiment of the entire Corps. The Corps meet was never held but this Regiment was selected by General Lewis to represent the 30th Division. General J. J. Pershing reviewed the Division while it was located in this area and later addressed the officers, expressing his appreciation for the work it had done while a member of the American Expeditionary Forces. A rating of the units of the Division was made by one of General Pershing's Staff Officers and shown to Colonel Herr, Chief of Staff, 30th Division. The 119th Infantry was rated number one in appearance, equipment and in marching in review.

"START FOR PORT."

On February 11, 1919, pursuant to orders, the Regiment proceeded by marching to the Forwarding Camps, five kilometers from LeMans, where it completed preparation before sailing, and on the twelfth and thirteenth days of March, proceeded to the Embarkation Camp at St. Nazaire, arriving March 13th and 14th. The Regiment remained at this camp for a period of three days, making its final preparations for embarkation to the United States of America.

On the 16th day of March, 1919, orders were received from Headquarters Embarkation Camp, Base Sector No. 1, St. Nazaire, France, to the effect that following units of the 119th Infantry would embark at 2:00 p. m., March 17th, on the U. S. S. "Madawaska".

Headquarters, 119th Infantry, Headquarters Company, Supply Company, and Ordnance Detachment, Machine Gun Company, 1st Battalion Headquarters, Companies A, B, C, and D, and Medical Detachment.

Following these units on March 21st, the 2nd and 3rd Battalion Headquarters, and Companies E, F, G, H, I, K, L, and M, embarked from St. Nazaire on the U. S. S. "Huron."

The trip over sea was very rough due to heavy wind and storms, thus causing the voyage to be several days longer than had been anticipated.

The entire Regiment landed at Charleston, S. C., on April 2nd, moving immediately by rail to Camp Jackson, S. C.

Here all the men in the Regiment from North Carolina, were mustered out of the service between the 6th and 10th, and those from Tennessee were sent to Fort Oglethorpe, Ga., April 9th, to be mustered out there. The remaining men in the Regiment from other States were transferred as Camp Detachments and ordered to the nearest Demobilization Camp to their homes and discharged. The following Camp Detachments were sent from the Regiment:

Camp Dix, Dodge No. 1 and No. 2, Gordon, Grant No. 1 and No. 2, Lee, Funston, Pike, Sherman, Taylor and Upton.

Practically all officers took advantage of the fifteen-day leave in order to avail themselves of an opportunity to secure employment before being discharged from service.

Upon the completion of mustering out of the 119th Infantry, its Commanding Officer, having applied for service in the Regular Army, appeared before a board consisting of Regular Army officers and was recommended by said board. He was not, therefore, mustered out with the remainder of the officers.

While at Camp Jackson, upon request of the city of Columbia, the Regiment on April 5th, moved to that city by way of electric railway and paraded through the principal streets, receiving a joyous welcome "HOME".

COLEMAN B. CARNWAY,
833 Main St., Danville, Va.

GEORGE A. SHUFORD
Asheville, N. C.

COLONEL S. L. FAISON

Formerly Commanding General 60th Brigade.

NOTE:--The following were the total casualties of the Regiment in its several operations:

OFFICERS

	July 4th Sept. 5th	*Sept. 29th Oct. 1st*	*Oct. 7th Oct. 12th*	*Oct. 16th Oct. 20th*	*Total*
Killed	3	----	----	3	6
Died of Wounds	----	1	----	2	3
Severely Wounded	6	5	3	4	19
Slightly Wounded	1	4	4	1	10
Gassed	1	----	1	----	2
Taken Prisoners	----	4	----	----	4
Undetermined Wounded	----	----	----	----	----

ENLISTED MEN

	July 4th Sept. 5th	*Sept. 29th Oct. 1st*	*Oct. 7th Oct. 12th*	*Oct. 16th Oct. 20th*	*Total*
Killed	64	146	56	28	294
Died of Wounds	12	15	6	3	36
Severely Wounded	50	123	63	43	279
Slightly Wounded	112	310	106	144	672
Gassed	45	103	30	45	223
Taken Prisoners	2	33	1	----	36
Undetermined Wounded	2	146	7	7	162

Five (5) Enlisted Men died of Other Causes.

ROSTER ENLISTED PERSONNEL

Name	Rank	Company	Date Joined	Wounded or Sick	Date Ret'd
Abbott, Ibon P	Pvt	D		W— Oct. 17-18	
Abernathy, George L	Cook	D		S— Dec. 2-18	Dec. 16-18.
Abernathy, Thomas H	Sgt	B		W— Sept. 29-18	Nov. 11-18.
Abbott, Johnnie R	Pvt--1st Cl	A		S— Aug. 18-18	Dec. 2-18.
Abraham, Paul R	Pvt	K		S— Oct. 14-18	Nov. 11-18.
Abromovic, Stanley	Pvt	I		W— Aug. 29-18	
Absher, Benjamin B	Pvt	F			
Acheson, James	Pvt	B	Dec. 6-18		
Acree, Joel F	Sgt	Hq		S— Oct. 29-18	Dec. 10-18.
Adams, James C	Pvt	F		S— Dec. 9-18	Dec. 20-18.
Adams, James R	Cook	G			
Adams, Marion B	Pvt	Sup			
Adams, Haymond	Pvt	A		W— Sept. 3-18	Dec. 2-18.
Adams, Roy	Pvt	G			
Adams, Archie B	Pvt--1st Cl	B			
Adams, Earl W	Pvt	G		W— Sept. 28-18	
Adams, James I	Pvt	L		S— Oct. 30-18	
Adams, John D	Mech	K		S— Aug. 8-18	

Adams, Wiley L	Pvt	K		S— 29-18 (Dec.)	
Adcock, William L	Pvt	B	Dec. 6-18		
Adcock, William L	Sgt	Hq			
Adkins, Arthur	Wag	Sup			
Adkins, John W	Pvt	H		K— Sept. 29-18	
Adkinson, Joseph B	Sgt	C		W— Oct. 14-18	
Agee, Byrd	Cpl	G			
Agney, George E	Pvt	Hq			
Aguere, Raymond	Pvt	E	Dec. 6-18		
Ahlstrom, Carl W	Pvt	D	Dec. 6-18		
Aiken, Lee	Pvt	Sup			
Aiken, Willie D	Pvt	M			
Airola, James A	Pvt--1st Cl	M			
Akers, Edward L	Mech	M		S— Nov. 11-18	Dec. 14-18.
Alberty, George	Pvt	D	Dec. 6-18		
Alberston, Eddie	Pvt	M		W— Oct. 10-18	
Albritton, Knot N	Pvt	D	Dec. 6-18		
Albritton, Robert L	Cpl	D			
Alderman, Jacob R	Pvt	A		K— Sept. 29-18	
Alexander, Earl	Pvt	D	Dec. 6-18		

Alexander, James M	Sgt	A			
Alexander, Joseph R	Cook	B			
Alexander, William E	Pvt	H			
Aley, Charlie	Pvt	D	Dec. 6-18		
Alford, Hansel	Pvt	D	Dec. 6-18	S— Jan. 2-19	Jan. 6-19.
Albright, Floyd E	Mech	E			
Allen, Connie B	Pvt	E		W— Sept. 29-18	Dec. 2-18.
Allen, Charles C	Pvt	H		W— Oct. 10-18	Nov. 26-18.
Allen, Charles	Pvt	A	Deserted Mar 16 1919.		
Allen, Claude W	Pvt	Md			
Allen, David	Pvt	M			
Allen, George S	Pvt--1st Cl	B		S— Jan. 5-19	Jan 10-19.
Allen, Roy	Pvt	D	Dec. 6-18		
Allen, William H	Pvt	D	Dec. 6-18		
Allen, William R	Pvt--1st Cl	E		S— Nov. 22-18	Dec. 22-18.
Allen, Irving	Pvt	D			
Allen, John Q	Sgt	M		K— Oct. 10-18	
Allesandre, Angelo	Pvt	D	Dec. 6-18		
Alligood, Elijah	Pvt	E			

Allison, Thomas J	Pvt	D	Dec. 6-18		
Allison, Arthur T	Pvt--1st Cl	G		K— Sept. 29-18	
Allred, Charlie L	Pvt	B		W— Oct. 17-18	
Allman, Fred R	Pvt	Hq		S— Nov. 26-18	Dec. 1-18.
Allman, John W	Pvt	D	Dec. 6-18		
Allman, John W	Pvt	K		W— Sept. 29-18	Dec. 5-18.
Allman, Seby	Pvt	D	Dec. 6-18		
Alphin, Raymond	Pvt	K		S— Jul 18-18	Sept. 15-18.
Althol, Ludwig C	Pvt--1st Cl	D		S— Oct. 14-18	Dec. 7-18.
Ambrose, Richard C	Pvt--1st Cl	B		T— Jul 6-18	
Amerson, Charles N	Cpl	L			
Amick, Walter	Mus-3d Cl	Hq			
Almond, Green W	Pvt	Sup			
Ammons, Clyde R	Pvt--1st Cl	Hq			
Amoia, Mikel	Pvt	B			
Anderson, Albert B	Pvt	C		K— Jul 21-18	
Anderson, Albert D	Pvt	F		T— May 4-18	
Anderson, E	Pvt	F		S— Oct. 1-18	
Anderson, James A	Pvt--1st Cl	Hq		T— Jan. 20-19	
Anderson, James D	Cpl	K		W— Aug. 4-18	

Anderson, Roy A	Pvt	A		W— Sept. 29-18	
Anderson, Charles	Pvt	D			
Anderson, Charlie S	Pvt	Hq			
Anderson, David	Pvt	M		W— Sept. 29-18 W— Oct. 10-18	Dec. 2-18.
Anderson, Edgar H	Pvt--1st Cl	B			
Anderson, George E	Pvt	H			
Anderson, George W	Pvt--1st Cl	K			
Anderson, John B	Pvt	D	Dec. 6-18		
Anderson, John E	Pvt	A			
Anderson, Lessie R	Pvt	C		W— Sept. 29-18	Dec. 29-18.
Anderson, Oather M	Pvt	Hq		T— Nov. 2-18	
Anderson, Robert	Pvt--1st Cl	C			
Andrews, Henry P	Sgt	I			
Andrews, Robert L	Pvt--1st Cl	E		K— Sept. 29-18	
Angle, Robert	Pvt	D		W— Sept. 29-18	
Anthony, Amzi	Cpl	H		S— Oct. 14-18 W— Sept. 29-18	Nov. 11-18.
Apoian, Hrand	Pvt	D	Dec. 6-18		
Archer, Vernie	Sgt	M			

Archuleta, Felimon	Pvt--1st Cl	B	Dec. 6-18		
Archero, Antonio	Pvt	D	Dec. 6-18		
Arms, Fowler	Pvt	B		W— Sept. 29-18	Dec. 10-18.
Armstrong, Chester B	Pvt	D		W— Jul 21-18	Sept. 15-18.
Armstrong, Isaac L	Pvt	MG		W— Sept. 29-18	Dec. 2-18.
Ambruster, John T	Pvt	M		W— Oct. 10-18	
Armstrong, Roger L	Cpl	B		W— Sept. 29-18	
Arnett, George W	Cpl	M			
Arney, Ernest G	Pvt	I		S— Oct. 3-18	Nov. 26-18.
Arnold, George H	Pvt	E		S— Dec. 19-18	Jan. 28-19.
Arnold, Floyd L	Pvt--1st Cl	B		W— Sept. 29-18	
Arp, Ford W	Pvt--1st Cl	Hq			
Arellano, Bemico	Pvt	E	Dec. 6-18		
Aricaetto, Donato	Pvt	A	Dec. 6-18		
Arminio, John	Pvt	E	Dec. 6-18		
Ash, Benjamin H	Cpl	D		K— Oct. 10-18	
Ashburn, William A	Pvt--1st Cl	L			
Ashe, Warren W	Pvt	F			
Askew, James E	Pvt--1st Cl	Hq			

Askew, Thomas J	Mech	B			
Askew, William B	Pvt	F		S— Aug. 18-18	Jan. 5-19.
Atchison, James S	Pvt	E	Dec. 6-18		
Atkins, Mack D	Cpl	G		S— Sept. 11-18	Dec. 2-18.
Atkinson, Arthur I	Pvt	E	Dec. 6-18		
Atkinson, Robert	Pvt--1st Cl	E			
Atry, Elias	Pvt	E	Dec. 6-18		
Aubrey, Johnnie S	Pvt	D		W— Sept. 29-18	Dec. 14-18.
Aukland, Jens	Pvt	D		W— Sept. 29-18	Dec. 7-18.
Aukes, Helmuth H	Cpl	I		D of W— Oct. 19-18	
Auld, William	Pvt	I		W— Aug. 31-18	
Autry, Lewis S	Pvt--1st Cl	L		W— Sept. 29-18	
Aulston, Clyde N	Sgt	Sup			
Ausburn, Dewey	Pvt	Hq			
Austin, Clyde L	Pvt	Hq			
Austin, Robert L	Pvt	A			
Austin, William H	Pvt	A		W— Oct. 17-18	Dec. 2-18.
Autry, Harry L	Pvt--1st Cl	L			
Avery, James B	Pvt--1st Cl	K			
Avery, Luther W	Pvt	M			

Avant, John R	Pvt	M		K— Sept. 29-18	
Avery, Jesse	Pvt--1st Cl	M		K— Aug. 4-18	
Aycock, William O	Sgt	K		W— Sept. 29-18	
Aycock, Willie	Pvt--1st Cl	E		S— Nov. 22-18	
Aydlette, Thomas F	Cpl	D			
Babb, Oscar T	Sgt	E		W— Sept. 29-18	Oct. 20-18.
Bachelor, Clender C	Pvt	A		S— Sept. 28-18	
Baddeur, Seff C	Sgt	Hq			
Bahuski, Wladislaw	Pvt	E	Dec. 6-18		
Bail, Oren B	Pvt	G		W— Oct. 10-18	Nov. 26-18.
Bailey, David E	Cpl	A		W— Oct. 10-18	Jan. 18-18.
Bailey, James E	Cpl	H		W— Oct. 16-18	Dec. 21-18.
Bailey, John A	Cpl	F			
Bailey, Oscar	Pvt	I			
Baird, Paul H	Pvt--1st Cl	E			
Baker, Clarence	Pvt	E	Dec. 6-18		
Baker, Conrad G	Pvt--1st Cl	D	Dec. 8-18.	W— Jul 20-18 W— Oct. 10-18	
Baker, George W	Pvt	E	Dec. 6-18		
Baker, Robert L	Cpl	A			

Baldwin, Arlie S	Wag	Sup			
Baldwin, Ed	Pvt	Hq			
Ballard, Edgar L	Cpl	B		S— Oct. 29-18	Nov. 27-18.
Ballard, Willie S	Pvt--1st Cl	MG			
Baltzli, John E	Pvt	B	Dec. 6-18		
Banke, Anthony F	Cpl	B			
Barbaria, Claudy	Pvt	E	Dec. 6-18		
Barbieri, Luigi	Pvt	E	Dec. 6-18		
Barbour, Lawrence D	Sgt	M			
Barbour, Norwood D	Mess. Sgt	K			
Barbrey, Frank	Pvt--1st Cl	MG		W— Sept. 29-18	Dec. 9-18.
Bardin, Oscar Y	Pvt--1st Cl	MG			
Barefoot, Zelotes	Sgt	M			
Barfield, Troy	Cpl	H		W— Oct. 10-18	Nov. 11-18.
Bartman, Albert W	Pvt	I			
Barletta, Steve	Pvt	E			
Barlow, James R	Pvt	A		S--	Dec. 14-18.
Barlow, William T	Pvt	E	Dec. 6-18		
Barnes, Dewey F	Pvt--1st Cl	G			
Barnes, Ellis	Pvt	B			

Barnes, John H	Pvt--1st Cl	K		W— Oct. 10-18	Dec. 20-18.
Barnes, Morris E	Mech	D			
Barnes, Wiley L	Sgt	Hq			
Barnett, Frederick A	Cpl	L		S— Nov. 11-18	Nov. 26-18.
Barnett, William	Pvt--1st Cl	M			
Barrett, James L	Pvt--1st Cl	G		W— Sept. 29-18	Dec. 2-18.
Barnwell, George	Pvt--1st Cl	G			
Barret, Albert	Pvt	I		W— Sept. 29-18	Dec. 5-18.
Barrix, Stokey A	Pvt--1st Cl	H		W— Jul 28-18	Oct. 27-18.
Barrow, Clyde R	Sgt	M			
Barrus, David R	Sgt	A			
Barry, Andrew J	Cpl	C		W— Oct. 10-18	Dec. 14-18.
Bart, Enrice	Pvt	E			
Bartlett, Alfred N	Pvt	Hq			
Barton, Burr	Pvt--1st Cl	F			
Bass, A. C	Cpl	Sup			
Bass, Lonnie E	Sgt	A			
Bass, Thomas B	Pvt	D			
Bassett, Orville L	Pvt	E	Dec. 6-18		

Bateman, Hershell	Pvt--1st Cl	Hq			
Batten, Hal	Pvt--1st Cl	MG			
Battisti, Frank R	Cpl	A			
Battle, James S	Sgt	MG		S--	Jan. 14-19.
Baty, Charles N	Sgt	L			
Bauer, John A	Pvt	B		S— Oct. 29-18	Nov. 26-18.
Baumgartner, Grover	Cpl	H			
Baumgartner, Fred	Pvt	D		W— Sept. 29-18	Nov. 26-18.
Baxley, James L	Cpl	L			
Beadles, William M	Pvt	A			
Bailey, Henry S	Pvt	H		W— Oct. 10-18	
Baird, Raymond E	Pvt	F		Deserted Dec. 10-18	
Baker, Carlos	Pvt	E		Died March, 4-19	
Baker, Isaac H	Pvt	K		S— Sept. 10-18	
Baker, Joe W	Cpl	L		T— Feb. 8-19	
Baker, Silas S	Pvt	G		W— Sept. 28-18	
Baldwin, George R	Pvt	H		K--Oct. 10-18	
Ball, John H	Pvt	A		K— Sept. 29-18	
Ball, Ralph B	Pvt	E		W— Oct. 10-18!	

Bancom, John A	Pvt	D		W— Sept. 29-18	
Barbour, William C	Pvt--1st Cl	C		Died— Oct. 10-18	
Barker, Clare D	Pvt	G		S— Oct. 1-18	
Barkley, John A	Sgt	D		S— Sept. 5-18	
Barnett, Frank N	Pvt	Hq		W— Oct. 17-18	
Basden, Ira	Pvt	B		W— Oct. 17-18	
Bass, Fred	Cpl	I		K— Sept. 29-18	
Bass, George W	Pvt	C		T— May 4-18	
Batson, Lewis E	Pvt	H		T— Jul 6-18	
Battle, Joseph D	Sgt	MG		T— Jan. 29-19	
Bauer, Fred E	Pvt	I		W— Sept. 29-18	
Baughan, Russel R	Pvt	A		W— Sept. 29-18	
Bavas, Hick A	Pvt	F		W— Sept. 29-18	
Beaman, William	Pvt--1st Cl	B		W— Oct. 17-17 T— Feb. 4-19	Oct. 17-18.
Beck, Russell	Pvt	G		W— Sept. 2-18 T— Jan. 22-19	Sept. 2-18.
Beck, William R	Pvt	M		W— Oct. 9-18	
Becker, Casper	Pvt--1st Cl	H		W— Sept. 29-18	
Becker, Victor	Pvt	I		K— Sept 28-18	

Beckman, John W	Pvt	A		W— Oct. 17-18	
Beckwith, Bryan	1st Sgt	F		T— Sept. 10-18	
Beeler, Adrian	Pvt	L		T— Jul 12-18	
Beland, Charles F	Pvt--1st Cl	K		T— Jun 19-18	
Belote, Emmett N	Pvt	A		S— Nov. 11-18	
Benson, Dennie A	Pvt	C		W— Jul 24-18 S— Mar 15-19	
Bently, John L	Pvt	Inf		K— Sept. 1-18	
Benton, Alonzo L	Pvt--1st Cl	A		W— Jul 28-18 T— Aug. 6-18	Jul 28-18.
Benton, Robert	Pvt--1st Cl	K		W— Sept. 29-18	
Berk, Max D	Pvt	G		T— May 4-18	
Betty, Wayne S	Pvt	B		W— Oct. 17-18	
Bevins, Thomas F	Pvt	E		K— Aug. 20-18	
Beake, Gowey L	Pvt	I			
Beamon, Charles F	Sgt	MG			
Beam, James S	Wag	Sup			
Bean, Auddie I	Pvt	E	Dec. 6-18		
Bean, Thomas C	Pvt	Hq			
Beard, Borden S	Sgt	L		W— Oct. 17-18	Nov. 26-18.

Beard, Silas	Wag	Sup			
Beard, Walter B	Pvt	I		W— Sept. 29-18	Dec. 14-18.
Beasley, William T	Pvt	Inf			
Beaty, Leslie	Pvt--1st Cl	G		S— Dec. 2-18	Dec. 7-18.
Beaver, Clifford T	Pvt--1st Cl	M			
Beck, John T	Wag	Sup			
Beck, Lawrence L	Pvt	A			
Beckwith, George	Mech	A			
Bedsloe, Ernest L	Sgt	F		W— Sept. 29-18	Dec. 2-18.
Bee, Gus	Pvt	A			
Beehler, John W	Pvt--1st Cl	C			
Behrends, Martin E	Pvt	K		W— Aug. 3-18 W— Sept. 29-18	Nov. 26-18.
Beighley, Morton E	Pvt	M		W— Oct. 10-18	Dec. 9-18.
Belk, Charles O	Cpl	F			
Bell, Charlie	Pvt	E	Dec. 6-18	S— Jan. 6-19	Jan. 15-19.
Bell, Chester	Pvt	E			
Bell, Clifton L	Pvt--1st Cl	E			
Bell, John T	Pvt	E	Dec. 6-18		
Bell, Ruben R	Ck	C			

Belli, Ignazio	Pvt	E	Dec. 6-18		
Belt, John L	Pvt	E	Dec. 6-18		
Benavides, Pablo E	Pvt	E	Dec. 6-18		
Bennett, Asa A	Pvt	A			
Bennett, Dove	Pvt--1st Cl	M			
Bennett, James C	Cpl	A		W— Sept. 29-18	Dec. 19-18.
Bennett, Sam M	Pvt	E	Dec. 6-18		
Bennett, Walter C	Sgt	M		S— Oct. 14-18	Dec. 14-18.
Bennett, Will C	Pvt--1st Cl	K			
Bennett, William J	Pvt--1st Cl	M		W— Oct. 10-18	Dec 14-18.
Bennett, Willie	Pvt--1st Cl	G		W— Sept. 29-18	Dec. 2-18.
Benson, Harry B	Pvt	E		W— Oct. 10-18	Dec. 11-18.
Benson, Victor E	Cpl	M		T— Oct. 31-18	Dec. 3-18.
Bentley, Oliver	Pvt	E	Dec. 6-18		
Benton, Jacob	Pvt	E	Dec. 6-18		
Benton, Lewis J	Sgt	Sup			
Benton, Louraine P	Sgt	C		S— Sept. 11-18	Dec. 2-18.
Benton, Luther M	Pvt	A		W— Aug. 17-18 W— Oct. 17-18	Dec. 2-18.
Berringer, John W	Pvt	E	Dec. 6-13		

Bernardini, Joe	Pvt--1st Cl	C			
Berry, John J	Pvt--1st Cl	Inf			
Best, James H	Sup Sgt	MG			
Bethune, Alexander S	BSM	Hq			
Bethune, James E	Mech	L			
Betro, Joe	Pvt	B	Dec. 6-18		
Betts, Willis A	Pvt	G			
Betty, Joseph H	Pvt	E		S— Oct. 31-18	Dec. 5-18.
Betty, Lincoln A	Pvt	I			
Beyers, Lester M	Mech	G		W— Sept. 29-18	Dec. 2-18.
Bickford, Carl B	Pvt	E	Dec. 6-18		
Bielefeld, Albert H	Pvt	I			
Bilbrey, Herschel V	Cpl	B		W— Sept. 29-18	
Billings, Garrett H	Sgt	Hq			
Bingham, Andrew O	Pvt	B		S— Aug. 26-18	Aug. 31-18.
Binka, John	Pvt	E	Dec. 6-18	S— Dec. 19-18	Dec. 28-18.
Binkley, Claude A	Pvt--1st Cl	C			
Birinske, Mike	Pvt	B	Dec. 6-18		
Bishop, Roy	Pvt	E	Dec. 6-18	S— Jan. 11-19	Jan. 16-19 .

Bistiritsky, Morris	Pvt	B	Dec. 6-18		
Bitner, Fred. S	Pvt	E	Dec. 6-18		
Bivins, Warren	Pvt	Hq		W— Sept. 29-18	Sept. 29-18.
Bjudstad, Albert	Pvt	E	Dec. 6-18		
Black, George M	Pvt	I			
Black, William A	Pvt	E	Dec. 6-18		
Biby, Robert L	Pvt--1st Cl	M		S— Oct. 20-18	
Biggers, Tollie M	Pvt	G		W— Jul 29-18	
Biggs, Harry	Pvt	B		T— Jul 6-18	
Biller, Edwin	Pvt	I		W— Oct. 17-18	
Bishop, James A	Pvt--1st Cl	F		K— Oct. 17-18	
Bissett, Bud	Pvt	Hq		S— Mar 16-19	
Bissett, Charles T	Sgt	A		T— Jul 22-18	
Black, James	Wag	Sup		S— Mar 15-18	
Blackman, Walter M	Mech	C		T— Jul 12-18	
Blackwell, Robert L	Pvt	K		K— Oct. 10-18	
Blake, Eugene	Pvt--1st Cl	H		W— Oct. 10-18	
Blalock, Dyar C	Pvt	K		W— Sept. 29-18	
Blalock, Luther L	Pvt--1st Cl	Hq		W- Sept. 29-18	

Blanchard, Leon C	Pvt	C		S— Mar 15-19	
Blankenship, Charles W	Pvt--1st Cl	L		S— Oct. 20-18	
Blevins, Bruce L	Cpl	D		K— Oct. 18-18	
Blevins, Walter	Pvt--1st Cl	F		W— Sept. 29-18	
Blair, James H	Pvt	E	Dec. 6-18		
Blake, Alva	Pvt	G			
Blakeley, Rippey J	Cpl	K		W— Sept. 29-18	Nov. 20-18.
Blaker, Russell	Pvt	M			
Blalock, Albert L	Pvt--1st Cl	B		W— Oct. 17-18	Dec. 2-18.
Blanchard, James H	Cpl	A		W— Oct. 17-18	Nov. 26-18.
Blanchard, Nathan M	Pvt--1st Cl	H			
Bland, Becton	Pvt	B			
Bland, Miller M	Pvt	E	Dec. 6-18		
Blanton, John C	Pvt	E	Dec. 6-18		
Blodgett, Alvin R	Pvt	E	Dec. 6-18		
Blue, Daniel A	Pvt	I			
Blue, Kenneth L	Cpl	Sup			
Boaz, Floyd	Pvt	L		S— Oct. 30-18	Dec. 21-18.
Bobbitt, Burnice B	Pvt	Hq			
Boehner, George	Pvt	C			

Boettcher, Henry A	Pvt	M		W— Sept. 29-18	Nov. 16-18.
Bogdeviez, Anthony	Pvt	E	Dec. 6-18		
Bogueki, John	Pvt	B			
Bolden, Rollie	Pvt--1st Cl	C		W— Sept. 29-18	Nov. 2-18.
Bolick, Odis S	Pvt--1st Cl	M		S— Oct. 1-18 S— Dec. 31-18	Jan 2-19.
Bolin, Bass C	Pvt	E	Dec. 6-18		
Bolls, Harvey W	Pvt	E	Dec. 6-18		
Bolt, Robert O	Pvt	MG			
Bond, John M	Cpl	L			
Bondi, Joseph	Pvt	E	Dec. 6-18		
Bonhardt, William H	Pvt	M		S— Sept. 29-18	Dec. 2-18.
Bonnell, David E	Pvt	A			
Bonnett, Louis H	Pvt	E	Dec. 6-18		
Boone, Chester	Pvt	M			
Boone, John E	Pvt--1st Cl	MG			
Boone, Lawrence J	Pvt	B		S— Oct. 29-18	Nov. 27-18.
Boone, Preston T	M Sgt	H			
Bordeaux, Anthony	Pvt--1st Cl	F		S— Oct. 20-18	Dec. 11-18.
Bost, Charlie R	Pvt	K		S— Oct. 29-18	Nov. 11-18.

Bost, J. Coyt	Pvt--1st Cl	G			
Bost, Walter C	Pvt--1st Cl	D		S— Oct. 28-18	Dec. 22-18.
Bostain, Gerthrie D	Wag	Sup			
Boswell, Clemson	Pvt	Sup			
Boswell, Thomas J	Sgt	I			
Botts, Allen	RSm	Hq			
Bowden, Ralph	Pvt--1st Cl	B			
Bowen, Connie	Pvt	M			
Bowen, Jesse L	Pvt	C		W— Sept. 29-18	Dec. 7-18.
Bower, Edgar J	Pvt--1st Cl	C		W— Oct. 17-18	Dec. 7-18.
Bowes, Ruben W	Pvt	K			
Bowman, James S	Pvt	A			
Bowles, Roy J	Pvt	E		S— Nov. 11-18	Dec. 11-18.
Boyd, Ernest J	Ck	Hq			
Boyd, James W	Pvt--1st Cl	E		S— Aug. 14-18	Dec. 2-18.
Boyd, Jesse A	Pvt--1st Cl	F			
Boyd, Jesse G	Pvt--1st Cl	L			
Boyd, William A	Pvt	L		W— Oct. 17-18	Dec. 2-18.
Boydston, Edgar M	Pvt	E	Dec. 6-18		
Boyette, Nathan	Wag	Sup			

Boyette, Neverson G	Sgt	B			
Boykin, John T	Pvt	D			
Boykin, Richmond T	Wag	Sup			
Boykin, Wade O	Pvt	Sup			
Boylan, Rufus T	M Sgt	MG			
Bozarth, Charles W	Sgt	Inf			
Brackett, James D	Pvt	Sup			
Bobst, Glen W	Pvt--1st Cl	Inf		T— Jan. 20-18	
Bodsk, Joseph	Pvt	I		S— Sept. 10-18	
Bodenberg, Albert C	Pvt	C		W— Oct. 17-18	
Boger, Charles I	Cpl	Hq		W— Aug. 19-18	
Boggess, Sidney E	Pvt	A		W— Oct. 17-18	
Boggs, Charlie	Pvt	D		W— Aug. 30-18	
Bolen, Ernest	Pvt	MG		S— Dec. 2-18	
Bolling, William	Pvt	A		K— Oct. 17-18	
Bond, Edward G	1st Sgt	L		T— Jul 22-18	
Bond, Gaither	Pvt	G		T— Jan. 22-19	
Bonner, Joseph N	Pvt	I		W— Oct. 10-18	
Boone, Jacob O	Cpl	A		W—Died Aug. 17-18	
Boone, Teddy M	Pvt	A		W— Sept 29.-18	

Bossert, Harry T	Pvt	C		W— Oct. 17-18	
Boswell, Ernest F	Pvt	K		S— Oct. 14-18	
Boulis, Peter G	Pvt	M		S— May 1-18	
Bovendar, Coleman W	Pvt	C		S— Oct. 14-18	
Bowen, William G	Pvt--1st Cl	L		W— Oct. 17-18 T— Feb. 8-19	Nov. 27-18.
Bower, Grover L	Pvt	MG		W— Sept. 27-18	
Bowers, John M	Cpl	F		S— Oct. 1-18	
Bowling, William T	Pvt--1st Cl	Hq		W— Oct. 17-18	
Boyd, David H	Pvt	E		K— Sept. 29-18	
Boyette, Kirby M	Pvt	D		W— Oct. 17-18 T— Jan. 20-19	Nov. 20-18.
Boyles, Roy M	Pvt	F		W— Sept. 29-18	
Brackett, Chauncey C	Pvt	A		W— Sept. 29-18	
Bradfield, Charles	Pvt	E		W— Oct. 17-18	
Brady, Ernest O	Pvt	Hq		W— Sept. 4-18	
Brady, Quitman	Pvt	B		W— Sept. 29-18	
Brand, John M	Ck	A		T— Jan. 29-19	
Brauer, Samuel C	Pvt	M		T— Sept. 2-18	
Bray, Charles E	Pvt	F		W— Sept. 29-18	

Breidenbaugh, Arlo H	Pvt--1st Cl	C		W— Jul 18-18	
Bremer, Herman F	Sgt	MG		K— Sept. 29-18	
Brewer, Frank J	Pvt	F		W--Died Sept. 29-18	
Brewer, Joseph S	Pvt	D		S— May 1-18	
Brickwell, William W	Pvt	E		W--Died Oct. 17-18	
Briedenbach, John J	Pvt	M		K— Oct. 10-18	
Brickley, Francis C	Pvt	H		S— Aug. 31-18	
Briggs, Ernest G	Cpl	F		W— Sept. 29-18	
Brigman, Harvey	Pvt	L		T— Feb. 8-19	
Brigmon, Arthur F	Pvt	H		W— Aug. 20-18 S— Oct. 1-18	
Briley, Jasper	Pvt	A		K— Oct. 17-18	
Brinson, Alonzo H	Pvt	D		S— May 1-18	
Brinson, Lee T	Cpl	B		W— Sept. 29-18	
Briody, Charles R	Pvt	C		S— May 1-18	
Britt, Jarvis L	Pvt	E		K— Sept. 29-18	
Britton, Rassie L	Pvt	E		S— Aug. 23-18	
Brehanner, Lawrence T	Pvt	I		W--Died Oct. 25-18	
Brooks, Isaac E	Sgt	Hq		T- Aug. 31-18	
Brooks, Leslie F	Pvt	G		S— Oct. 29-18	

Brown, Alvin C	Pvt	C		S— Dec. 12-18	
Brown, Archie S	Pvt	A		S--Died Oct. 14-18	
Brown, Claude J	Pvt	D		W— Jul 25-18 T— Jul 27-18	Jul 25-18.
Brown, David B	Cpl	C		W— Oct. 17-18 T— Jan. 20-19	
Brown, Edgar	Pvt	G		S— Aug. 1-18	
Brown, Edgar H	Pvt--1st Cl	D		S— May 1-18	
Brown, Erman R	Pvt	A		W— Sept. 29-18	
Brown, Floyd B	Pvt--1st Cl	C		W— Oct. 10-18	
Brown, George M	Pvt--1st Cl	F		W— Sept. 29-18	
Brown, Harry	Pvt	A		W— Sept. 29-18	
Brown, John	Pvt	E		W— Aug. 23-18 W— Sept. 23-18 T— Jan. 22-19	
Brown, Lex C	Pvt	K		W--Died Oct. 11-18	
Brown, Rockwell V	Ck	Sup		T— Jan. 20-19	
Brown, Zeb	Mech	D		S— Sept. 19-18	
Brownstein, Sam	Pvt	I		Deserted Oct. 1-18	
Broyles, Samuel K	Pvt	Inf		T— Oct. 28-18	

Brunett, Oscar	Cpl	G		S— Sept. 15-18	
Brune, Sarafine	Pvt	F		S— May 1-18	
Bruno, Dominick	Pvt--1st Cl	M		S— Oct. 20-18	
Bruns, Herman T	Pvt	M		W— Aug. 5-18	
Brush, Wesley	Pvt	D		K— Sept. 29-18	
Bryant, Alvin E	Cpl	K		W— Sept. 29-18	
Bryson, Henry L	Pvt--1st Cl	K		K— Oct. 16-18	
Bryson, Thomas N	Pvt--1st Cl	G		W— Sept. 29-18	
Bryson, Walter E	Pvt	K		S— Aug. 12-18	
Brady, Carlton	Pvt--1st Cl	D		W— Aug. 20-18	Dec. 2-18.
Brady, Richard D	Sgt	D		S— Oct. 28-18	Dec. 11-18.
Brady, Robert L	Pvt	B			
Brafford, Lawrence W	Pvt--1st Cl	D		W— Sept. 29-18	Nov. 26-18.
Brandenburg, Henry	Cpl	F		W— Aug. 25-18 W— Sept. 29-18	Oct. 29-18.
Brannon, Vergil T	Pvt	E	Dec. 6-18	S— Jan. 7-19	Jan. 16-19.
Brantley, Elijah D	Pvt	Hq			
Bravard, Charles H	Pvt	C			
Brazeel, Helmer H	Pvt	E	Dec. 6-18	S— Jan. 6-19	Jan. 15-19.
Breekenridge, Chester A	Sgt	K			

Breeze, John A	Cpl	H			
Breihan, Carl C	Pvt	F	Dec. 6-18		
Brendle, Lester	Cpl	A			
Brendle, William	Pvt	G			
Brewer, Archie R	Pvt	F	Dec. 6-18		
Brewer, Robert B	Mech	F			
Brewer, William E	BCpl	Hq			
Brezenski, William A	Pvt	F	Dec. 6-18		
Bridges, Boyd	Pvt	M			
Bridges, William L	Pvt	F		W— Aug. 18-18 W— Sept. 29-18	Nov. 6-18.
Bridgewater, Harvey	Pvt	H		W— Oct. 10-18	Dec. 2-18.
Brinkley, Boney J	Pvt	F	Dec. 6-18		
Brinkley, Herbert C	Pvt	K			
Brinkley, John R	Mech	K		W— Sept. 29-18	Nov. 26-18.
Bristow, Bernie N	Sgt	G			
Brite, Freddie	Pvt	I			
Britt, David E	Pvt	G			
Britt, James H	Sgt	I			
Britt, John W	Pvt--1st Cl	Ordn			

Britt, Robert E	Pvt	A			
Britt, Samuel M	RSM	Hq			
Brittain, Harry A	Pvt	F	Dec. 6-18		
Brittain, Stanley	Pvt--1st Cl	D		S— Nov. 11-18	Nov. 26-18.
Brittner, Daniel	Pvt	A		W— Aug. 31-18	Dec. 11-18.
Broadwater, Paul M	Pvt	A			
Broadway, Walter R	Mus	Hq			
Brock, Horace F	Cpl	E		W— Oct. 17-18	Dec. 11-18.
Brock, Thomas B	Pvt	F	Dec. 6-18		
Bronski, Julius	Pvt	F	Dec. 6-18		
Brooks, James B	Cpl	K			
Brooks, Thomas M	Cpl	I			
Broone, Edward C	Pvt	D		W— Oct. 17-18	Dec. 7-18.
Broughton, Charlie	Pvt	Sup			
Brown, Arthur F	Pvt	E			
Brown, Artice C	Pvt	H			
Brown, Claude H	Sgt	D			
Brown, Daniel L	Pvt	F	Dec. 6-18		
Brown, Eazor F	Pvt--1st Cl	H		W— Sept. 29-18	Dec. 19-18.

Brown, Elmer E	Pvt--1st Cl	C			
Brown, Fred A	Cpl	C		W— Sept. 1-18	Sept. 1-18.
Brown, Frederick E	Pvt	F			
Brown, George F	Pvt	A		W— Sept. 29-18	Dec 2-18.
Brown, Huriah P	Pvt	K			
Brown, James B	Pvt	F	Dec. 6-18		
Brown, James M	Cpl	C			
Brown, John T	Pvt	A		S— Oct. 19-18	Dec. 2-18.
Brown, Lemuel F	Pvt	C			
Brown, Lewis	Pvt	M			
Brown, Max	Pvt--1st Cl	MG			
Brown, Norman E	Pvt	F	Dec. 6-18		
Brown, Oakes	Pvt	F	Dec. 6-18		
Brown, Ollie T	Pvt--1st Cl	C			
Brown, Oscar H	Sgt	I		W— Sept. 29-18	Dec. 14-18.
Brown, Owen L	Pvt	F		W— Sept. 29-18	Dec. 2-18.
Brown, Ray T	Pvt	C			
Brown, Roy J	Mus	Hq			
Brown, Weldon	Pvt--1st Cl	C		S— Oct. 29-18	Nov. 8-18 .

Brunntta, Tony	Pvt	B	Dec. 6-18		
Bryan, William J	Pvt	F	Dec. 6-18		
Bryant, Charles	Pvt	K			
Bryant, Clarence R	Cpl	I			
Bryant, James A	Pvt--1st Cl	Hq			
Bryant, Willie T	Pvt	Hq			
Bryson, Hardie C	Cpl	I		W— Aug. 23-18	Aug. 23-18.
Brozzowski, Peter	Pvt	F	Dec. 6-18		
Buchanan, George W	Pvt	G		W— Sept. 3-18 W— Sept. 29-18 Missing in action	Oct. 10-18.
Buff, Robert L	Pvt	F		S— Jul 6-18	
Bullard, Frank R	Cpl	I		W— Oct. 10-18	
Bullock, Ernest	Pvt	G		S— May 1-18	
Burcham, McKinley	Pvt	I		S— Aug 11-18!	
Burgess, Urious I	Pvt--1st Cl	G		S--Died Sept. 25-18	
Burnette, Charles H	Sgt	F		T— Jul 22-18	
Burns, James B	Pvt	C		S— Sept. 25-18	
Burns, James E	Pvt	F		W— Sept. 29-18	
Burris, James B	Pvt	A		W— Oct. 17-18	

Butler, Clarence	Sgt	K		K— Oct. 10-18	
Butler, Haywood	Cpl	M		T— Jan. 16-19	
Byerley, Alva E	Cpl	C		W— Aug. 31-18	
Byram, Leslie A	Pvt	A		W— Oct. 17-18	
Byrd, Akrie J	Cpl	M		K— Oct. 10-18	
Byrd, Zeke	Cpl	H		S— Oct. 31-18	
Buchanan, Burlan C	Pvt	C			
Buchanan, Grady D	Pvt	K			
Buchanan, James E	Cpl	Hq			
Buchanan, Rufus	Pvt--1st Cl	I		W— Sept. 29-18	Dec. 11-18.
Buchan, Thomas	Pvt	H			
Buck, Charles A	Pvt	F	Dec. 6-18		
Buck, James A	Pvt--1st Cl	B		S— Oct. 29-18	Nov. 26-18.
Buck, Robert	Pvt--1st Cl	A		W— Sept. 29-18	Nov. 11-18.
Buckner, Herschell J	Pvt	D			
Budzinski, Francis	Pvt	B	Dec. 6-18		
Bufford, Warren	Cpl	L			
Bugliono, Gennaro	Pvt	B	Dec. 6-18		
Bulich, Bob	Pvt	G	Dec. 6-18		

Bull, John M., Jr	Pvt--1st Cl	Inf		S— Oct. 1-18	Dec. 30-18.
Bullard, James J	Sgt	F		W— Sept. 29-18	Nov. 11-18.
Bu'lock, Noah M	Pvt--1st Cl	K		W— Sept. 29-18	Jan. 18-19.
Bumpus, Andrew H	Sgt	L		W— Oct. 17-18	Dec. 14-18.
Bunch, Richard W	Sgt	I			
Bunn, Joseph P	1st Sgt	MG		S— Nov. 8-18	Nov. 14-18.
Bunn, Percy L	Ck	MG			
Burgbower, Cornelius	Pvt	A			
Burgess, Henry L	Pvt	G	Dec. 6-18		
Burkey, Edward R	Pvt	I			
Burneikis, Selvester	Pvt	MG			
Burnett, William P	Sgt	E			
Burnette, Jason L	Pvt	D		S— Oct. 1-18	Nov. 26-18.
Burnette, Miley	Pvt	F			
Burnette, Wates	Wag	Sup			
Burroughs, John H	Pvt	G	Dec. 6-18		
Burrows, Rufus B	Pvt	Hq	Dec. 6-18		
Bursell, George	Pvt	C			
Burt, John	Pvt	I			

Burton, Floyd G	Pvt	G	Dec. 6-18		
Bush, Bazzle	Pvt	D			
Bushnell, Charles C	Pvt	H		S— Oct. 31-18	Nov. 14-18.
Bussie, Pete	Pvt	I		M— Sept. 29-18	Nov. 11-18.
Butler, Cliff	Pvt	H		M— Sept. 29-18	Nov. 26-18.
Butler, Emory L	Pvt--1st Cl	K			
Butler, Harvey K	Pvt	F			
Butler, Jesse D	Pvt	B			
Butler, John R	Wag	Sup			
Butler, Oscar L	Pvt	I		W— Sept. 29-18	Dec. 14-18.
Butler, Robert H	Pvt	H			
Butters, Ulysses E	Pvt	A			
Butts, Zeb V	BCpl	Hq			
Byrd, Fred A	Cpl	L		S— Nov. 11-18	Mar 14-19
Byrne, Harry T	Pvt	E			
Byrum, Lyn S	Sgt	I			
Cagle, Charles L	Pvt	K			
Cade, Jay	Pvt	D		W— Oct. 17-18	
Cagle, Fred. C	Pvt--1st Cl	D		W— Sept. 29-18	Dec. 29-18.

Caffey, Rone L	Pvt	D		K— Oct. 18-18	
Canola, Joe	Pvt	B		S— Nov. 11-18	Dec. 2-18.
Cahoon, Lonnie L	Pvt--1st Cl	C		W— Oct. 17-18	
Cain, Cleve	Pvt	Hq			
Calaway, Robert A	Pvt	C		W— Oct. 17-18	Oct. 17-18.
Caldwell, Harrison	Pvt	F		W— Oct. 10-18	Nov. 26-18.
Caldwell, Grover C	Pvt	K		T— June 19-18	
Callahan, Lyman T	Wag	Hq			
Callahan, Henry	Ck	E			
Calloway, Zeb L	Sgt	M			
Campbell, Ellis	Pvt	G	Dec. 6-18		
Calvert, Eliza C	Pvt	L		S— Aug. 3-18	
Calvert, Harry H	Pvt--1st Cl	I		K— Sept. 29-18	
Campbell, June M	Pvt	Hq		T— Dec. 6-18	
Campfield, Jack	Pvt	E		W— Sept. 29-18	
Campbell, Frank T	Pvt--1st Cl	E		W— Sept 29-18	Dec. 14-18.
Canady, Hugh	Pvt	K			
Canady, Alvin	Cpl	I		Died of wounds Aug 2-18.	
Canady, Leon G	Cpl	H			
Canale, James M	Pvt--1st Cl	C		S— Aug. 15-18	Dec. 2-18.

Cane, Bert D	Pvt	E		W S— Oct 20-18	
Cannon, Claude	Pvt	G		K— Sept. 29-18	
Cancilla, Frank	Pvt	G	Dec. 6-18		
Cannon, Daniel	Corp	G			
Cannon, Frank	Pvt	G		K— Sept. 29-18	
Cantelmo, Gaetno	Pvt	G	Dec. 6-18		
Canter, Charlie G	Pvt--1st Cl	D			
Cantey, John J	Sgt	Sup			
Capps, Ransom	Pvt--1st Cl	A			
Cantrell, Radford	Pvt--1st Cl	B		W— Sept. 29-18	
Capps, Ernest F	Pvt--1st Cl	H		W— Oct. 10-18	
Carabello, Demenic	Pvt	G	Dec. 6-18		
Caramezza, Louis	Pvt	G	Dec. 6-18		
Carvaun, Freeman C	Pvt	C		W— Sept. 29-18	Dec. 7-18.
Carawan, Guy H	Corp	MG		S— Aug. 26-18	Aug. 29-18.
Carbough, John S	Pvt	G	Dec. 6-18		
Carden, Ed	Pvt--1st Cl	H		W— Oct. 10-18	Oct. 25-18.
Carguino, Martino	Pvt	C		W— Oct. 17-18	
Carlisle, Perry G	Pvt--1st Cl	I		S— Oct. 31-18	Dec. 22-18 .

Carlisle, Thomas W	Sgt	D			
Carlisle, John W	Pvt--1st Cl	K		Died of disease Feb. 16-19.	
Carlson, Noble G	Pvt	I		W- -Sept. 29-18	Oct. 10-18.
Carlson, Carl J	Pvt	D		W— Sept. 29-18	
Carlson, Hilmer T	Pvt	F		W— Sept. 29-18	
Carlson, Ervin G	Pvt	D		S— Nov. 11-18	
Carlton, John H	Pvt--1st Cl	B		W—S est. 29-18	Dec. 1-18.
Carpenter, Ellis H	Pvt	A			
Carpenter, Oscar H	Pvt	I			
Carnes, Eddie M	Pvt	D		W— Oct. 17-18	
Carr, Ezra	Cpl	B		W— Oct. 17-18	Nov. 27-18.
Carringtor, Robert F	Pvt	M		S— Dec. 16-18	Jan. 8-18.
Carson, Campbell	Pvt	L			
Carson, William B	Sgt	A			
Carson, William F	Pvt	Sup			
Carter, John S	Pvt	H			
Carter, William H	Cpl	L			
Carothers, John W	Pvt	C		W— Sept. 29-18	
Carpenter, Charlie	Pvt	L		T— Jan. 29-19	

Carpenter, Loyd W	Pvt	M		K— Oct. 11-18	
Carpenter, Schyler McK	Pvt	E		W— Sept. 29-18	
Carson, Victor H	Pvt	C		S— Oct. 15-18	
Carter, Ben F	Pvt	L		K— Sept. 2-18	
Carter, Buck A	Pvt--1st Cl	H		K— Sept. 29-18	
Carter, Dalton D	Pvt--1st Cl	H		W— Sept. 29-18	
Carter, James A	Pvt	E		S— Oct. 29-18	
Carter, Gene	Pvt	D		W— Oct. 17-18	
Carter, Raymond C	Pvt	G		T— Sept. 2-18	
Cartwright, George P	Pvt	F		T— May 4-18	
Cartwright, Hillery	Pvt	I			
Cartwright, John	Pvt	H		W— Sept. 29-18	Dec. 19-18.
Cartwright, Sim E	Pvt--1st Cl	E			
Caruso, Domenick	Pvt	G	Dec. 6-18		
Carver, John	Cpl	H		W— Oct. 10-18	Nov. 11-18.
Casey, Edward J	Pvt--1st Cl	M		S— Nov. 20-18	Dec. 21-18.
Casey, Ernest R	Cpl	Hq		T— Mar 15-19	
Casey, Guy F	Pvt	A		S— Nov. 15-18	
Casey, Luther R	Pvt--1st Cl	B		W— Sept. 29-18	

Cash, Sherman S	Cpl	D		W— Oct. 17-18	Nov. 26-18.
Castet, Fernand, Jr	Pvt	A			
Cassiola, Sam M	Cpl	C		Died of wounds Oct. 13-18.	
Cates, Herbert	Pvt	F		K— Sept. 29-18	
Cates, Jones A	Mech	H		S— Oct. 28-18	Nov. 26-18.
Cates, William S	Cpl	K			
Cattaellino, Guiseppe	Pvt	G	Dec. 6-18		
Caudill, James E	Pvt--1st Cl	G		K— Sept. 29-18	
Caudle, Dallas	Pvt--1st Cl	E		W— Oct. 17-18	Nov. 3-18.
Causey, Benjamin L	Cpl	D			
Cavanaugh, Macon E	Pvt--1st Cl	MD			
Cavanaugh, Robert R	Pvt	C		W— Oct. 17-18	Jan. 3-19.
Caviness, Dewey	Pvt	D		W— Sept. 29-18	Nov. 11-18.
Caviness, Joseph E	Cpl	L		W— Oct. 10-18	
Caynor, Harley	Pvt	A		S— Nov. 15-18	Dec. 8-18.
Caywood, Edgar E	Pvt	M		S— Oct. 31-18	
Corato, Jim	Pvt	L		W— Aug. 31-18	
Chaffin, Raymond R	Pvt	E		W— Sept. 29-18	
Chamberlain, Victor P	B.S. M	Hq		T— Jan. 1-19	

Chambers, Loyd	Pvt	MG		W— Sept. 29-18	
Chandler, John R	Pvt--1st Cl	B			
Chance, Ernest	Pvt--1st Cl	E		W— Aug. 20-18	Oct. 15-18.
Chance, Jasper	Pvt--1st Cl	E		W— Jul 30-18	Sept. 23-18.
Chandler, Pleasant O	Pvt--1st Cl	C		W— Oct. 29-18	Jan. 9-19.
Chart, Alfred H	Pvt	G	Dec. 6-18		
Chase, John W	Pvt	Hq	Dec. 6-18		
Chase, Marion B	Pvt	Sup			
Chapman, Clifton	Pvt	A		K— Oct. 10-18	
Cheatam, Clyde	Pvt--1st Cl	Hq		T— Aug. 10-18	
Chatelain, Victor J	Pvt	M			
Cheuhar, Anthony	Pvt	G	Dec. 6-18		
Cherry, George E	Pvt--1st Cl	C			
Chennis, Joseph W	Cpl	I		T— Jan. 29-19	
Cherry, Stone F	Pvt	B		K— Sept. 29-18	
Cheslock, Michael J	Pvt	I		S— Nov. 27-18	
Chesteen, Ernest	Pvt	M		S— Oct. 20-18	
Charella, William	Pvt	D			
Chaitti, Amaton	Pvt	G	Dec. 6-18		

Chism, Clay	Pvt	G	Dec. 6-18		
Christensen, Christian	Cpl	K			
Chesson, James H	Cpl	Sup			
Christoff, John J	Pvt	MG			
Chipps, Stanley D	Pvt	A		W— Oct. 17-18	
Charhart, Edward J	Pvt--1st Cl	E		Died of wounds Jul 19-18.	
Christie, Charles O	Pvt	MD		T— Nov. 20-18	
Church, Clinnard	Pvt	H		W— Sept. 29-18	Nov. 11-18.
Church, Henry	Pvt	G		T— Aug. 25-18	
Church, Lindsay J	Pvt--1st Cl	D		W— Sept. 29-18	
Church, Lewis S	Mech	B			
Cieri, Angelo	Pvt	B	Dec. 6-18		
Cierlicki, Frank	Pvt	G	Dec. 6-18		
Ciernia, Nicholas	Pvt	C			
Cifers, James E	Pvt	F		W— Sept. 29-18	
Civils, Bennie O	Pvt	F		K— Sept. 29-18	
Clark, Charles E	Pvt--1st Cl	C			
Clark, Curtis L	Pvt	G			
Clark, Elijah H	Pvt	MG			

Clark, Isaac O	Pvt	Hq			
Clark, John R	Cpl	F		T— Oct. 28-18	
Clark, James	Pvt--1st Cl	E			
Clark, Ralph L	Pvt--1st Cl	Hq		W— Aug. 4-18	Dec. 5-18.
Clark, William A	Pvt--1st Cl	E		W— Oct. 19-18	Oct. 20-18.
Clanton, Richard	Pvt	H		K— Oct. 17-18	
Clark, Dougall	Pvt	I		W— Sept 29-18	
Clark, George S	Cpl	K		W— Oct. 10-18	
Clark, Sidney C	Pvt	B		W— Oct. 17-18	
Clark, Victor W	Pvt	Hq		S— Oct. 1-18	
Clark, Winnie D	Pvt	D		W— Sept. 29-18	
Clayton, John T	Mech	I		S— Oct. 31-18	
Clayton, Roy C	Pvt--1st Cl	E		T— Jul 12-18	
Claxton, Wesley R	Cook	L			
Clay, John G	Cook	H			
Clery, William H	Pvt	Hq			
Clever, Charlie C	Sgt	Sup			
Clement, Stephen A	Sgt	A		W— Sept. 29-18	Dec. 19-18.
Clemmer, Will R	Pvt--1st Cl	C			

Clemmons, Ernest	Pvt	C			
Clevenger, Carl	Bug	A			
Clemmons, Clamer T	Cpl	F		W— Sept. 29-18 W— Oct. 18-18	
Clifton, Ivory S	Pvt	F		T— May 4-18	
Clifton, Willie E	Sgt	M		S— Oct. 14-18	Dec. 2-18.
Clymer, Flavie V	Pvt	H			
Cline, Albert W	Pvt	Hq			
Cline, Elmer W	Sgt	H			
Cline, Lawrence R	Pvt	H		W— Sept. 29-18	Dec. 19-18.
Cline, Carnell S	Sgt	G		W— Aug. 20-18	
Cline, Frank E	Pvt	I		Died of wounds Sept. 29-18.	
Clippard, Klutz	Cpl	H		W— Sept. 29-18	
Clontz, Charlie F	Pvt--1st Cl	C			
Clow, Earl E	Pvt--1st Cl	H		W— Sept. 29-18	Nov. 11-18.
Clowers, Geroge E	Pvt--1st Cl	Hq			
Clowes, Myrr	Pvt	Hq			
Clute, John B	Pvt--1st Cl	MG			
Coakley, Harvey	Pvt--1st Cl	E			
Coan, Frank	Pvt	A		S— Nov. 15-18	Nov. 27-18.

Cobasco, John A	Pvt	I		S— Sept. 13-18	Oct. 27-18.
Cobb, Claude S	Pvt	MG		S— Jan. 1-19	Jan. 9-19.
Cobb, Carlisle A	Pvt--1st Cl	D		K— Sept. 29-18	
Cobb, Glenn H	Pvt	I		S- Aug. 21-18	Dec. 11-18.
Cobb, Robert A	Corp	I		S— Jul 6-18	Oct. 27-18.
Cochran, Charles B	Pvt	B		W— Sept. 26-18	Dec. 8-18.
Cochran, David R	Pvt	I		K— Oct. 7-18	
Cochran, Jesse C	Wag	Sup			
Coff, Major	Bug	B		W— Oct. 10-18	
Cockman, Isem W.	Pvt	I			
Cockrell, Walter W	Pvt	F		W— Sept. 29-18	Nov. 11-18.
Coe, Joseph C	S-Sgt	G		W— Sept. 29-18	Dec. 19-18.
Coe, Wilson V	Corp	F		W— Sept. 29-18	Nov. 11-18.
Coffman, Rollo A	Pvt	C		W— Sept. 29-18	Dec. 7-18.
Coggins, James C	Sgt	I		T— Jul 22-18	
Colmanti, Guiseppe	Pvt	F		W— Oct. 10-18	
Cole, Floyd H	Pvt	Hq		S— Sept. 6-18	Dec. 1-18.
Cole, Mack	Pvt	B		W— Oct. 10-18	Dec. 11-18.
Coleman, Avery L	Corp	D			

Coleman, Charles	Pvt--1st Cl	E		W— Oct. 10-18	Dec. 2-18.
Coleman, Charlie W	Corp	A		S— Oct. 18-18	Nov. 1-18.
Coleman, Ernest	Sgt	M		S— Nov. 27-18	Dec. 10-18.
Coleman, Fred D	Pvt	M		K— Oct. 11-18	
Coleman, Junius P	Pvt--1st Cl	Hq		T— Sept. 22-18!	
Coleman, James P	Ck	Hq			
Coleman, John F	Pvt--1st Cl	B		W— Oct. 17-18	Dec. 2-18.
Coleman, Peter J	Pvt	F		T— Jan. 20-19	
Coleman, William R	Corp	D		W— Oct. 17-18	
Colin, Earl E	Pvt	F		W— Oct. 10-18	Dec. 11-18.
Coll, John F	Pvt	F		W— Oct. 10-18	
Collier, Ewing H	Corp	E			
Collins, James C	Pvt	G		W— Oct. 10-18	Dec. 11-18.
Collins, Luther J	Pvt	B		W— Sept. 26-18	
Collins, Neal N	Pvt	Hq		W— Sept. 29-18	
Collins, Sanford	Pvt	B			
Collins, Walter M	Pvt	G		S— Oct. 1-18	
Colucci, Cosimo	Pvt	G	Dec. 6-18		
Colucci, Giovanni	Pvt	G	Dec. 6-18		

Colvert, William T	Pvt	E			
Coomey, Aloysius B	Pvt--1st Cl	T		K – Oct. 18-18	
Comarata, Marine	Pvt	E		S— Oct. 29-18	Nov. 27-19.
Combs, Levi	Wag	Sup			
Comer, Lawrence	Pvt	F		W— Oct. 10-18	Dec. 10-18.
Comers, Connie L	Pvt--1st Cl	H		W— Sept. 29-18	
Compton, Charles C	Pvt	A			
Conatzer, Clay	Corp	G			
Cornatser, Emit	Corp	B		W— Oct. 10-18	Dec. 1-18.
Conaway, Nason F	Pvt	B		Died of W--Sept. 29-18.	
Conder, Harry W	Pvt--1st Cl	D			
Conger, Francis W	Pvt--1st Cl	L			
Conley, Mark	Pvt	G		W— Sept. 29-18	Nov. 26-18.
Conley, Robert L	Pvt--1st Cl	G		W— Aug. 20-18 W— Sept. 29-18	Dec. 2-18.
Conley, Willie P	Pvt--1st Cl	G		S— Oct. 29-18	Dec. 14-18.
Conlin, Arthur	Pvt--1st Cl	B		W— Oct. 17-18	Dec. 14-18.
Connell, John	Pvt	F			
Conner, Claude R	Pvt--1st Cl	B		S— Nov. 11-18	Dec. 21-18.
Conrad, Woodfron G	Pvt	H		W— Oct. 10-18	Nov. 26-18.

Conrad, Oscar J	Pvt	F		S— Oct. 29-18	
Comron, William P	Wag	Sup			
Conway, William A	Pvt	F			
Cook, Henry E	Pvt--1st Cl	C		W— Oct. 17-18	Dec. 2-18.
Cook, Lawrence S	Pvt--1st Cl	K		S— Nov. 26-18	Dec. 22-18.
Cook, Isom W	Pvt	I		T— Feb. 12-19	
Cook, Lee	Pvt	G		W— Oct. 10-18	
Cook, Lloyd H	Sgt	L		W— Aug. 31-18	
Cook, Preston J	Pvt--1st Cl	L		W— Oct. 10-18	
Cook, Walter W	Corp	F		S— Aug. 31-18	
Cooke, Claude	Pvt	A		W— Sept. 29-18 T— Jan. 29-19	Jan. 3-19.
Cooke, George C	Pvt	Hq		T— Jan. 20-19	
Cooney, Aloysius B	Pvt--1st Cl	C		K— Oct. 18-18	
Cooper, Talmage	Pvt	E	W--Sept. 29-18		
Cooper, William N	Pvt	L		K— Oct. 11-18	
Cooper, Bally	Pvt	K		W— Sept. 29-18	Oct. 14-18.
Cooper, Bennie A	Pvt	B			
Cooper, Joseph L	Corp	C			
Coppedge, Edward S	Pvt--1st Cl	L		W— Sept. 29-18	Dec. 22-18.

Copeland, William A	Pvt	MD		T— Sept. 1-18	
Copeland, Alex	Pvt	Hq		T— May 4-18	
Corbett, James E	Pvt	A		S— Nov. 27-18	
Corbett, Tudor	Pvt	B		T— Jul 6-18	
Corbett, Belzeni A	Sgt	C			
Corbett, William H	Corp	C			
Cordell, John R	Mech	C		W— Aug. 22-18 T— Jan. 28-19	Sept. 1-18.
Cornwell, Herber F	Pvt	E		W— Sept. 29-18	
Cordon, William W	Pvt--1st Cl	E			
Cornatzer, George F	Wag	Sup			
Corrigan, John	Pvt	MG	Oct. 26-18		
Coryna, Ira H	Pvt	E		W— Sept. 29-18	
Coulter, George N	Pvt	E		S— Oct. 1-18	
Councill, Bernard B	Corp	A		Died of W--Oct. 9-18	
Couzins, Leroy	Pvt--1st Cl	K		W— Sept. 29-18	
Coston, Luther	Pvt--1st Cl	M		S— Aug. 6-18	Aug. 16-18.
Cotton, Thomas H	Pvt--1st Cl	K		W— Sept. 29-18	Nov. 11-18.
Couch, Jethro C	Pvt	H		S— Oct. 29-18	Dec. 7-18.

Covington, Newton J	Ck	Hq			
Cox, Ed	Pvt	G		K— Sept. 29-18	
Cox, Roland S	Pvt	MG		K— Sept. 29-18	
Cowan, Edward	Pvt--1st Cl	B			
Cox, Charles C	Pvt--1st Cl	Hq		S— Nov. 1-18	Dec. 5-18.
Cox, Emmett H	Pvt	Sup			
Cox, John F	Pvt	E			
Cox, Willie E	Corp	C		S— Sept. 11-18	Sept. 21-18.
Cox, Willie J	Corp	M			
Craig, James H	Pvt	E			
Crabtree, Joe W	Pvt--1st Cl	Hq		S— Jan. 3-19	
Craft, Walter A	Pvt	C		S— Sept. 28-18	
Crane, Nathan E	Pvt	E		T— May 4-18	
Crane, Robert M	Pvt	H		W— Sept. 29-18	
Cranford, Jesse T	Pvt	F		W— Oct. 10-18	
Cramer, Loyd K	Pvt	G	Dec. 6-18		
Cranfill, Harvey L	Corp	D		W— Oct. 10-18	Dec. 7-18.
Cratty, Thomas	Pvt	G	Dec. 6-18		
Crawford, Edward L	Pvt	Sup			

Crawford, Rufus H	Pvt--1st Cl	C			
Crawford, William E	Sgt	C		S— Oct. 28-18	
Crawshaw, Gilbert	Pvt--1st Cl	B		W— Oct. 17-18	
Crayton, Charles	Pvt	E			
Creasy, Woodie	Pvt	A			
Creekmore, Beecham	Pvt	E			
Credle, Daniel C	Pvt	MG		W— Oct. 17-18	
Crews, William	Pvt	K		K— Oct. 9-18	
Creason, Edward	Pvt--1st Cl	E		W— Sept. 23-18	Dec. -18.
Cress, Gustave H	Corp	M			
Crews, Elijah	Pvt--1st Cl	B			
Crisp, Albert D	Pvt	Sup			
Crisp, Loyd E	Pvt	C			
Criswell, Earl F	Pvt	I		S— Oct. 31-18	Nov. 11-18.
Criswell, Shuford C	Pvt	MD			
Critchelow, James	Pvt	MD			
Crites, Manuel H	Pvt	C		S— Oct. 28-18	Dec. 14-18.
Crocker, John H	Pvt--1st Cl	C			
Crockett, Do k	Pvt	G		S— Oct. 29-18	Dec. 2-18.

Crockett, Peyton	Pvt	D		T— June 21-18	
Cronick, Joseph	Pvt	F		W— Sept. 29-18	
Crook, Bert	Pvt	I		S— Aug. 1-18	
Croom, William J	Sgt	A		T— Jan. 29-18	
Cronin, Dennis A	Pvt	C			
Croom, Clyde	HS	Sup			
Crouch, Worth B	Corp	L			
Crouch, Barney E	Pvt--1st Cl	G		K— Sept. 1-18	
Crowder, Thomas	Pvt--1st Cl	B		S— Oct. 28-18	
Crow, Ute	Pvt--1st Cl	H		W— Oct. 10-18	Oct. 27-18.
Crowder, Manlie G	Pvt	I		S— Oct. 31-18	Nov. 11-18.
Crum, Benjamin J	Pvt	F	Oct. 26-18		
Crum, Edward	Pvt	C		S— Oct. 29-18	Dec. 7-18.
Crumpler, David D	Corp	H		S— Nov. 1-18	Dec. 2-18.
Crutchfield, William B	Pvt	F		W— Sept. 9-18	
Crusco, Vito	Pvt	G	Dec. 6-18		
Crutchfield, William M	Pvt	G		S— Jan. 7-19	Jan. 15-18.
Cudsworth, Joseph H	Pvt	E		W— Oct. 10-18	
Cummings, J. B	Pvt	B		W— Sept. 29-18	

Cummings, Anthony A	Corp	B			
Cummings, William J	Sgt	E		W— Oct. 10-18	Oct. 29-18.
Cunningham, Ben	Pvt	B		K— Oct. 19-18	
Cunningham, Willie	Pvt--1st Cl	E		W— Sept. 29-18	
Cunningham, Wisie P	Pvt--1st Cl	E		K— Sept. 29-18	
Cunningham, Chas P	Pvt	MG		W— Oct. 10-18	Nov. 14-18.
Cunningham, French H	Pvt	A			
Cunningham, Lewis F	Pvt	B	Dec. 6-18		
Curat, George J	Pvt	G	Dec. 6-18		
Curry, Claude B	Pvt	G	Dec. 6-18		
Curtis, Benjamin L	Pvt	C		W— Oct. 17-18	Dec. 11-18.
Cushman, Walter A	Pvt	Hq		W— Oct. 17-18	Dec. 29-18.
Cusimano, Nicola	Pvt	G	Dec. 6-18		
Cutler, Augustus	Pvt	I		S— Oct. 21-18	Dec. 2-18.
Cutter, Walter	Pvt	E		W— Sept. 29-18	
D'Abbieri, Nicola	Pvt	G		W— Oct. 10-18	Dec. 14-18.
Dackis, Charles A	Pvt--1st Cl	F		W— Oct. 10-18	
Dahlinghaus, August J	Pvt--1st Cl	L			
Dale, Edward T	Cpl	C		S— Aug. 10-18	Nov. 11-18.

Dalton, Samuel	Pvt	MG		S— Dec. 2-18	
Danford, Henry F	Cook	I			
Daniels, David	Pvt	Sup			
Daniels, Monroe A	Pvt--1st Cl	E			
Daniels, William A	Pvt	E			
Daniels, William H	Pvt	G			
Daniels, Add E	Pvt	MG			
Daniels, James C	Pvt	E		W— Sept. 29-18	Dec. 11-18.
Daniels, Joseph C	Pvt--1st Cl	MG		W— Sept. 29-18	Dec. 14-18.
Daniels, Marion L	Sgt	E		K— Sept. 29-18	
Danksis, Kazis	Pvt	G		W— Sept. 29-18	Dec. 14-18.
Dannels, Jesse F	Pvt	Hq		W— Aug. 25-18 T— Jan. 20-18	Dec. 21-18.
Dannel, Tobias Q	Sgt	Hq		T— Jan. 20-18	
Dannen, John M	Pvt	F			
Dannenbring, Fred C	Pvt	K		W— Sept. 29-18	
Dansereau, Elmer J	Pvt	I		W— Oct. 10-18	
Dart, Percy T	Pvt--1st Cl	F		W— Sept. 29-18	
Darling, Fenny	Pvt	I			
Darrall, Curtis W	Pvt	C			

Daughterty, Roscoe T	Cpl. Bnd	Hq			
Daughterty, Charlie	Cpl	B		T— Feb. 4-18	
Daghtry, Leon L	Sgt	H		K— Sept. 29-18	
Davenport, Dennis W	Pvt	K		W— Aug. 25-18	
Davidson, Samuel	Pvt--1st Cl	F		W— Sept. 29-18 W— Oct. 10-18	Oct. 15-18.
Davidson, Starling	Pvt	C			
Davidson, Delbert	Pvt	D		W— Sept. 29-18	
Davidson, Henry M	Pvt	M		S— Oct. 1-18	
Davidson, William C	Pvt	G		K— Oct. 18-18	
Davis, Harry L	Pvt	C		T— Jul 6-18.	
Davis, Homer R	Pvt	Hq		T— Sept. 11-18	
Davis, John	Pvt	E		W— Oct. 10-18	
Davis, John F	Pvt--1st Cl	B		W— Sept. 29-18	
Davis, Omer	Pvt	C		K— Sept. 29-18	
Davis, Phillip	Pvt	C		K— Sept. 29-18	
Davis, Stanley	Pvt	G		W— Oct. 10-18	
Davis, Arthur H	Pvt	C			
Davis, Beverage	Pvt	G			
Davis, Carter L	Pvt	I		W— Sept. 29-18	Oct. 27-18.

Davis, Clement G	Cpl	C		W— Oct. 10-18	Dec. 2-18.
Davis, Dock	Cook	E			
Davis, Dunk	1st Sgt	G		W— Oct. 17-18	Nov. 11-18.
Davis, Edwin B	Sgt	MG			
Davis, George	Pvt--1st Cl	E			
Davis, Harry L	Pvt--1st Cl	MD		W— Jul 22-18	Aug. 1-18.
Davis, Harvey	Cpl	C			
Davis, James	Pvt	F			
Davis, James A	Wag	Sup			
Davis, James H	Bug	E		W— Sept. 29-18	Dec. 14-18.
Davis, Jesse S	Pvt	M			
Davis, Joe R	Pvt	E		W— Sept. 29-18 W— Oct. 17-18	Dec. 3-18.
Davis, John	Wag	Sup			
Davis, Joseph	Pvt--1st Cl	D			
Davis, Junius	Cpl	Hq			
Davis, Luke	Pvt--1st Cl	Hq		S— Dec. 7-18	Jan. 4-19.
Davis, Major	Cpl	G		W— Oct. 10-18	Oct. 27-18.
Davis, Marshall A	Cpl	L			
Davis, Richard A	Hs	Sup			

Davis, Rufus	Pvt	Hq			
Davis, Stephen R	Pvt	A			
Davis, William G	Pvt	I			
Davis, Willie L	Pvt--1st Cl	MG			
Davieau, Oscar F	Pvt--1st Cl	M			
Dayton, Leo	Pvt	G		S— Oct. 1-18	
Dean, Anthony F	Pvt	MG		Died of disease Feb. 27-18.	
Dawald, William L	Pvt	K			
Dawson, Amos M	Pvt	H		W— Sept. 29-18	Nov. 20-18.
Dawson, Chester	Pvt	C			
Dawson, Clarence W	Pvt--1st Cl	MG		W— Sept. 29-18	Oct. 15-18.
Dawson, Herman	Pvt--1st Cl	C			
Day, John R	Pvt	K		W— Sept. 29-18	Dec. 14-18.
Dayton, John C	Pvt--1st Cl	G			
Dean, David	Pvt	C		W— Aug. 31-18	
Dean, Walter C	Pvt	F			
DeAgelangelis, Gervasio	Pvt	G	Dec. 6-18		
Dearman, DeWitt T	Pvt	K		W— Sept. 29-18	Nov. 14-18.
DeBerry, George F	Pvt	E		W— Sept. 29-18	Nov. 26-18.

DeBoer, William	Pvt	E		W— Oct. 10-18	Nov. 26-18.
DeBrue, Horace	Pvt	Sup			
Dees, Dan	Pvt--1st Cl	Hq			
Deese, Joel S	Pvt	K		W— Aug. 26-18	Dec. 8-18.
Deesing, Peter S	Pvt	MG			
DeHaan, Henry J	Pvt	H			
Dehr, John A	Cpl	I			
Delvechio, Robert H	Pvt	D		S— Aug. 25-18	Dec. 22-18.
Dempsey, James C	Sgt	K		T— Jul 22-18	
Demsky, Joseph	Pvt	F		W— Oct. 29-18	
Denbo, Raleigh	Pvt	F		S— Oct. 29-18	
Dennis, Duffy	Cook	A		S— Sept. 28-18	
Dennis, Judson W	Sgt	L		K— Oct. 17-18	
Dennis, Robert	Cpl	D		T— Oct. 13-18	
Denny, Davis W	Cpl	L		W— Aug. 21-18	
Denton, Sib	Pvt--1st Cl	B		W— Oct. 28-18	
Deil, Charles E	Cpl	MG			
Deil, Jerry	Pvt	MG		S— Dec. 28-18	Dec. 31-18.
DeLaney, Hobard	Pvt	I			

Delduco, Generoso	Pvt	B	Dec. 6-18		
Deloatch, Willie L	Pvt	A			
Dennery, Patrick V	Pvt	MG			
Denning, James L	Pvt	B		S— Dec. 22-18	Jan. 2-18.
Dennis, Jasper	Cpl	A		W— Sept. 29-18	Nov. 1-18.
Dent, Sydney H	Pvt	C			
Dentici, Antonio	Pvt	G	Dec. 6-18		
Deoppe, John E	Pvt	C		S— Nov. 11-18	Dec. 12-18.
Derwin, Edward D	Pvt	E		T— May 4-18	Jan. 29-19.
Dew, Caldwell	Cpl	D		W— Sept. 29-18	
Dew, Harvey	Pvt--1st Cl	L		S— Sept. 10-18	Sept. 19-18.
Dewey, Frank	Bug	D		W— Oct. 17-18	
Deweese, Wesley	Pvt	M			
Dickens, Harper	Pvt				
Dickerson, Ernest	Pvt	H			
Dickerson, Carl L	Cpl	C		S— Oct. 28-18	
Dillard, Ben T	Sgt	Hq		W— Sept. 4-18	
Dillard, William E	Pvt--1st Cl	MD		T— Aug. 18-18	
Dillon, Loyd H	Pvt	G		K— Sept. 29-18	

Dietz, George H	Pvt	G			
Dietz, Harry F	Pvt	MG			
Dinges, Emil J	Pvt	M		T— June 22-18	
Dinopoli, Michael	Pvt	G	Dec. 6-18		
Dishop, Emil	Pvt	F			
Disney, Robert L	Pvt--1st Cl	MD			
Dixon, Clarence E	Pvt	A		W— Oct. 10-18	Nov. 25-18.
Dixon, John	Pvt	M		W— Oct. 10-18	Nov. 11-18.
Dixon, Drew E	Pvt--1st Cl	M D		T— Jan. 20--18	
Dixon, Clarence	Sgt	G		W— Sept. 29-18	
Dixon, George T	Pvt	E		K— Sept. 29-18	
Dixon, John R	Pvt--1st Cl	A		S— Nov. 15-18	
Dobson, Ralph J	Cpl	K		W— Sept. 29-18	Dec. 5-18.
Dobyss, Leo M	Cpl	C		S— Aug. 8-18	Dec. 2-18.
Dodge, William T	Pvt	F			
Dodd, Alonzo	Pvt--1st Cl	C		T— Jul. 15-18	
Doerr, Otto J	Pvt	I			
Dominuco, Vincenzo	Pvt	H		W— Sept. 29-18	Dec. 14-18.
Domikas, William	Pvt	L		W— Oct. 17-18	

Donald, Roy	Pvt	E		S— Mar 16-19	
Donaldson, Willis N	Cpl	M			
Donnan, Fred C	Pvt	M		S— Mar 16-18	
Donnerson, William V	Pvt--1st Cl	D		K— Oct. 10-18	
Doran, Carl T	Pvt--1st Cl	Hq		S— Sept. 10-18	Sept. 25-18.
Dorazio, Camille	Pvt	G	Dec. 6-18		
Dortch, Hugh	Sgt	MG			
Dorman, Jack J	Pvt	MG		S— Nov. 18-18	Nov. 30-18.
Dorsey, Lester L	Pvt	H		W— Sept. 29-18	Nov. 14-18.
Dugan, Thomas C	Pvt	G	Dec. 6-18		
Douglass, Ottis	Pvt--1st Cl	C		W— Sept. 29-18	
Dougherty, John T	Pvt	B		W— Sept. 29-18	Nov. 5-18.
Douglass, Andrew J	Pvt	MD			
Dowd, Ralph M	Pvt	Hq		W— Oct. 17-18	
Driver, James C	Cpl	D		K— Sept. 29-18	
Drach, Hugo A	Pvt	F		W— Oct. 10-18	
Drake, Herman H	Pvt--1st Cl	MG			
Dranth, Sam	Pvt--1st Cl	C		S— Oct 28-18	Nov. 26-18.
Drew, Charles A	Pvt	G		W— Sept. 29-18	Jan. 10-18.

Driggs, Otey T	Pvt	G		W— Sept. 29-18	No. 6-18.
Drinkard, Homer L	Cpl	Hq			
Driver, Roy H	Pvt	F		W— Sept. 29-18	Nov. 11-18.
Dudeck, Tony	Pvt	G	Dec. 6-18		
DuHamel, George	Pvt--1st Cl	L			
Duncan, Elmer C	Cpl	Hq		W— Sept. 28-18	Dec. 1-18.
Dummon, John R	Pvt	E		S— Nov. 2-18	Dec. 3-18.
Durham, Lester	Cpl	L			
Durham, Ransom C	Pvt	D		W— Sept. 29-18	Dec. 14-18.
D'Urso, Caesar	Pvt	G	Dec. 6-18		
Dush, Rae E	Pvt	Hq			
Duvall, Harry	Pvt	C			
Duvall, Howard	Pvt	Hq			
Drum, Will	Pvt	G		T— Jan. 21-19	
Dubberly, Jesse T	Pvt	E		W— Oct. 10-18	
Duddy, Peter J	Pvt	G		W— Oct. 18-18	
Duffy, Michael J	Pvt	G		W— Sept. 29-18	
Dugan, Thomas	Pvt	H		T— Jan. 20-19	
Dula, William S	Pvt--1st Cl	G		Sept. 29-18 T— Jan. 22-19	Oct 27-18.

Dunlap, Charlie W	Pvt	E		W— Sept. 29-18	
Dunlap, Fred	Pvt--1st Cl	G		T--Jan. 22-19	
Dunn, Frank	Cpl	H		K— Sept. 29-18	
Dutler, Emil H	Pvt	E		W— Oct. 17-18	
Dwen, Albert	Pvt	H		S— Oct. 31-18	
Dwyer, John C	Pvt	I			
Dyer, John R	Cpl	E		S— Sept. 6-18	Sept. 15-18
Dyer, William C	Pvt	G	Dec. 6-18		
Eads, Alva A	Pvt--1st Cl	B			
Eagan, Hugh S	Pvt--1st Cl	Hq			
Eagan, James J	Pvt--1st Cl	G		S— Oct. 14-18 T— Jan. 22-18	Dec. 7-18.
Eagan, Rufe	Pvt	E		S— Oct. 14-18	
Eagan, Walter J	Corp	I		W— Oct. 14-18	
Eaks, Barbee	Pvt	E		T— Jan. 22-18	
Eaks, Joe H	Pvt	E		S— Nov. 2-18	
Early, Rudolph	Pvt	G	Dec. 6-18	S— Dec. 24-18	Feb. 5-19.
Early, Henry G	Mus	Hq			
Earnhardt, Fred L	Pvt	A		W— Oct. 10-18	Nov. 27-19.
Eason, Otto	Pvt--1st Cl	G		W— Sept. 29-18	Dec. 2-18.

Eastep, Ralph	Sgt	H			
Easter, Clifton M	Pvt	Hq		S— Jul 28-18	Aug. 12-18.
Eckhert, Henry G	Pvt	I		S— Nov. 11-18	Nov. 20-18.
Edens, George T	Corp	G		S— Dec. 2-18	Dec. 5-18.
Ederhoff, George J	Pvt	G	Dec. 6-18		
Edge, Johnnie S	Pvt	Sup			
Edgeman, James H	Pvt	F			
Edwards, Arthur F	Pvt--1st Cl	B		S— Aug. 31-18	Dec. 19-18.
Edwards, Fred M	Pvt	G	Dec. 6-18		
Edwards, Herman	Pvt	L			
Edwards, John A	Pvt	F			
Edwards, John R	1st Sgt	Hq			
Edwards, Johnathan C	Pvt	Hq			
Edwards, Sam E	Pvt--1st Cl	E			
Edwards, Wayne S	Corp	D			
Edwards, Wayne R	Pvt	Hq		W— Aug. 19-18	Sept. 3-18.
Edwards, William E	Pvt	E			
Edwards, William W	Pvt	F			
Eden, Earl A	Pvt	I		S— Oct. 14-18	

Edge, Jasper	Mech	F		T— Jan. 20-19	
Edwards, Carsey C	Pvt	H		Died of W— Oct. 1-18	
Edwards, James J	Corp	C		W— Sept. 29-18	
Edwards, James T	Pvt	L		W— Oct. 17-18	
Edwards, William F	Pvt	E		K— Sept. 29-18	
Eikleberger, James	Pvt	F		W— Sept. 29-18	Dec. 4-18.
Eidson, Neuma R	Pvt	E			
Elder, Arza C	Pvt	M		W— Oct. 10-18	
Eller, James A	Pvt	G		K— Sept. 29-18	
Eller, Raleigh A	Pvt	MG		M— Sept. 29-18	
Elliott, Daisy	Pvt	L		S— Nov. 11-18 T— Feb. 8-19	Nov. 16-18.
Elliott, Floyd L	Pvt	B		S— Nov. 11-18	
Elliott, Raymond V	Pvt--1st Cl	F		K— Oct. 14-18	
Elliott, Seaf	Pvt	C		S— Jul 6-18	
Elliott, Frank C	Pvt	H		S— Oct. 31-18	Nov. 14-18.
Elliott, Joseph R	Pvt	C			
Elliott, Morris A	Corp	MG			
Ellis, Edwin D	Pvt--1st Cl	L			

Elliss, Haywood	Pvt--1st Cl	C			
Ellis, James	Pvt	C		W— Aug. 31-18 W— Oct. 17-18	Nov. 26-18.
Ellis, Ned	Ck	D			
Ellis, William M	Pvt--1st Cl	Hq			
Ellis, Joseph H	Pvt	MG		T— Jan. 28-19	
Ellis, Walter	Pvt	K		T— Jan. 19-18	
Ellis, Wess	Pvt	B		W— Sep. 26-18	
Elmore, George	Pvt	G		W— Oct. 17-18 T— Jan. 22-19	Nov. 16-18.
Elmore, Julius C	Pvt	G		W— Sept. 29-18	
Embler, McCoy L	Pvt	Hq			
Emmons, Ray W	Pvt	M			
Emolovich, Andrew	Pvt--1st Cl	B			
Emory, Furman M	Pvt	F		S— Mar 16-18	
Endres, Jacob C	Corp	H			
Endrizzi, John B	Pvt	H			
Englebard, Ervin G	Corp	G		K— Sept. 29-18	
Engle, Joseph C	Sgt	D		W— Sept. 29-18	Nov. 1-18.
Enlow, Edward A	Corp	I		W— Oct. 10-18	Dec. 2-18.

Ennis, Frank	Pvt	M		S— Nov. 11-18	Dec. 3-18.
Enoch, Abram	Pvt	B			
Enright, Bryan	Pvt	I		T— Feb. 20-19	
Epperson, Edward A	Pvt	G		K— Oct. 18-18	
Epstein, Leonard	Sgt	MG		T— Mar 16-19	
Epps, Herman L	Pvt	D		W— Sept. 3-18 T— Feb. 2-19	Sept. 3-18.
Ervin, Ernest M	Corp	K		K— Aug. 26-18	
Ervin, William D	Pvt	MD		T— Jan. 29-19	
Ergle, Ernest C	Pvt	F			
Ernest Herbert C	Pvt	Hq			
Erwood, Brenton C	Pvt	M		T— June 22-18	
Eslick, John H	Pvt	E		W— Sept. 29-18 T— Jan. 22-19	Dec. 19-18.
Estes, Ralph C	Sgt	Hq		T— Jul 22-18	
Estes, Ernest C	Sgt	B		T— Oct. 13-18	
Estep, Solomon W	Pvt	F		S— May 1-18	
Estep, Max	Pvt	MG			
Etheridge, Ernest T	Pvt	MD			
Etheridge, Walter	Pvt	M		W— Sept. 29-18	Jan. 18-19.

Etheridge, Paul	Pvt	Sup			
Ethier, Joseph A	Pvt	E		W— Oct. 14-18	Dec. 14-18.
Etter, Elbert C	Corp	Hq			
Etters, Arthur	Pvt	I		W— Aug. 31-18	
Eubank, Elmore E	Sgt	B		W— Sept. 26-18 W— Oct. 10-18	Nov. 8-18.
Eubanks, Verny G	Pvt	F		S— Nov. 11-18	
Eudy, Fred B	Pvt	I			
Eure, Solomon Q	Corp	F		W— Jul 27-18	Nov. 1-18.
Evans, Eugene M	Corp	MG			
Evans, James C	Pvt	L			
Evans, Joe H	Pvt	Hq		K— Aug. 18-18	
Evans, Joseph C	Pvt--1st Cl	K		K— Sept. 29-18	
Evans, Luther	Pvt	Hq			
Everett, Cecil L	Corp	H		W— Aug. 26-18 W— Sept. 29-18	Dec. 2-18.
Everhart, Avoy	Pvt--Ast Cl	D			
Everly, Guy E	Pvt	MG		W— Sept. 27-18	Oct. 10-18.
Evers, David J	Pvt	F			
Ewell, Jesse O	Ck	A			

Ewell, Charlie M	Pvt	D		T— Feb. 4-19	
Exum, George F	Pvt	K			
Ezzell, John R	Pvt	A		Died of W--Sept. 29-18	
Ezzell, David E	Pvt--1st Cl	MG			
Fabbri, Adelmo	Mech	L			
Fabisiak, John	Pvt--1st Cl	B			
Facclinno, Goivanni	Pvt	H		W— Oct. 10-18	Dec. 14-18
Fay, John J	Pvt	A			
Fair, Robert K	Pvt--1st Cl	D		K— Oct. 18-18	
Faircloth, Archie	Pvt	I		Died of W--Sept. 29-18	
Faircloth, Josiah S	Cpl	F		W— Jul 18-18	
Faircloth, Gibson	Pvt	F		W— Sept. 29-18	Nov. 6-18.
Faircloth, Henry C	Pvt--1st Cl	H		W— Oct. 10-18	Dec. 15-18.
Faircloth, Lamar	Pvt	A		W— Oct. 10-18	Nov. 26-18.
Faircloth, Lonzy	Pvt	F		W— Oct. 17-18	Nov. 26-18.
Fairless, Andrew J	Pvt	E			
Falk, James L	Sgt	L		S— Oct. 20-18	Dec. 2-18.
Falletti, Mike	Pvt	B	Dec. 6-18		
Farr, Harley	Pvt	Hq			

Farr, Joe	Pvt	Hq		S— Nov. 11-18	Dec. 15-18.
Farrar, Lewis P	Pvt	F			
Farris, Marion	Cpl	G		T— Jul 6-18	Aug. 10-18.
Fann, Lank	Cpl	M		T— Jan. 16-19	
Faulkner, Thomas V	Cpl	E		W— Oct. 10-18	
Fasig, Fred A	Pvt	Hq	Dec. 6-18		
Faucett, John	Pvt--1st Cl	Hq			
Faucett, Thomas C	Cpl	M			
Faulkner, Richard W	Ck	Hq			
Faulstich, Walter	Pvt--1st Cl	M			
Foust, Henry H	Pvt	A		W— Sept. 29-18	Nov. 11-18.
Feidak, Nuan	Pvt	G	Dec. 6-18		
Ferrell, Leroy	Pvt	B		S— Nov. 11-18	Nov. 27-18.
Ferguson, Guilford A	Sgt	Sup			
Ferguson, Lucas B	Pvt--1st Cl	MG		W— Sept. 29-18	Nov. 26-18.
Ferguson, William	Pvt	F		W— Sept. 29-18	
Ferguson, Roy	Pvt	E		W— Sept. 29-18	Dec. 7-18.
Ferrell, Thomas J	Pvt	F		T— Jan. 20-18	
Ferrell, William T	Pvt	H		W— Sept. 29-18	

Ferrantello, James	Pvt	G	Dec. 6-18		
Ferrell, Frank B	Sgt	K			
Ferrell, Seth	Cpl	K			
Fiasst, Hugo	Pvt	B	Dec. 6-18		
Fickey, Charlie E	Pvt--1st Cl	Hq			
Fickey, Harry H	Cpl	Hq		W— Sept. 29-18	Oct. 28-18.
Field, William J	1st Sgt	E		T— Jul 22-18	
Field, Ellery S	Pvt	MG		S— Oct. 7-18	
Fink, Ode	Pvt	H		T— Jul 6-18	
Fiderlick, Henry E	Cpl	D			
Fielden, Robert S	Pvt--1st Cl	L		W— Oct. 17-18	Nov. 26-18.
Fields, Charles	Pvt--1st Cl	I		T— Nov. 28-18	Dec. 3-18.
Fink, John H	Cpl	K			
Fink, Lucy	Cpl	M			
Fitch, Alonzo E	Pvt	I		K— Oct. 10-18	
Fitzgerald, William B	Cpl	L		W— Aug. 31-18 W— Sept. 29-18 T— Feb. 8-18	
Fitshugh, Arthur S	Pvt	L		T— Sept. 14-18	
Fitzsimmons, Robert J	Pvt	G		K— Oct. 16-18	

Fipps, Chess	Pvt	Hq			
Fischer, Fred W	Pvt	M		W— Oct. 10-18	Dec. 3-18.
Fischer, John	Pvt	E			
Fisher, Julius J	Pvt	MG		S— Nov. 8-18	Dec. 7-18.
Fischer, Leo G	Pvt	M			
Fischer, George H	Pvt	C			
Fisher, John L	Cpl	E		W— Sept. 29-18	Nov. 20-18.
Fisher, Samuel L	Sgt	C			
Flack, John T	Cpl	K			
Flack, Charles	Pvt	I		D of W— Sept. 2-18	
Flannigan, Andrew W	Pvt	E		T— May 4-18	
Flannigan, John	Pvt	I		W— Sept. 26-18	
Flaspohier, Frank	Pvt	F		W— Sept. 29-18	Dec. 3-18.
Flauaus, Alfred P	Pvt	E		T— Jul 12-18	
Fleetwood, Mallie	Pvt	I		W— Aug. 8-18	
Fiesher, Guy	Pvt	C		W— Oct. 17-18	
Fleming, Harold T	Pvt--1st Cl	A		W— Aug. 16-18	Nov. 1-18.
Fleming, Hiram B	Pvt	C		W— Oct. 17-18	Nov. 26-18.
Flesier, Manley	Pvt	H		W— Jul 26-18	Dec. 30-18.

Fletcher, Floyd R	Pvt	I			
Flickinger, Harry C	Pvt	MG			
Flick, John L	Pvt	I		S— Oct. 20-18	
Florengi, Louis	Pvt	G	Dec. 6-18		
Flora, James	Pvt	C		S— Nov. 11-18	
Flowers, Clem B	Mech	Hq			
Flowers, George J	Pvt	H		W— Sept. 29-18	Dec. 11-18.
Floyd, Roscoe	Pvt	C		W— Sept. 29-18	Dec. 7-18.
Fly, Thomas J	Sgt	A		W— Oct. 10-18	
Flynn, Homer	Pvt--1st Cl	M			
Flynt, Rabah B	Pvt--1st Cl	A		S— Oct. 29-18	Nov. 14-18.
Folger, Paul B	Cpl	MG		W— Sept. 29-18	Nov. 8-18.
Folks, George F	Pvt	G		W— Sept. 29-18	Dec. 14-18.
Follette, Clarence C	Pvt	Sup			
Folden, Homer H	Pvt	C		S— Oct. 28-18	
Foor, Harley J	Pvt	I			
Forbes, Bert T	Cpl	I			
Forbes, John S	Asst Bn Ldr	Hq			
Ford, James W	Pvt	G			

Ford, Jesse G	Pvt--1st Cl	L		K— Aug. 2-18	
Forrest, Samuel P	Pvt	G		S— Oct. 29-18	
Forhand, Walter S	Sgt	I		T— Oct. 13-18	Jan. 3-19.
Forland, Harold A	Cpl	A			
Foster, Walter I	Pvt	D		W— Sept. 29-18	
Foscue, Troy E	Cpl	Hq			
Foster, Ben W	Pvt	L		S— Dec. 1-18	Dec. 21-18.
Foster, Charles J	Pvt	Hq			
Foster, Ezra P	Pvt--1st Cl	A		W— Oct. 17-18	Dec. 2-18.
Foster, Luke	Pvt	B	Dec. 6-18		
Fowler, Troy	Pvt	A		S— Oct. 1-18	
Fouts, Rufus D	Pvt	L		S— Nov. 11-18	Nov. 26-18.
Fowler, Ransom D	Pvt	Sup			
Fowler, Thomas	Pvt	Sup			
Fowlkes, Dock F	Cpl	K		S— Nov. 29-18	Dec. 2-18.
Fox, Clarence L	Pvt	G	Dec. 6-18		
Fox, Jacob F	Pvt	F		K— Jul 31-18	
Fox, Manuel A	Pvt	C		W— Aug. 31-18	
Foxman, Benjamin	Bnd. Sgt	Hq		S— Nov. 26-18	Jan. 1-19.

Foy, Leslie	Pvt	H		K— Sept. 29-18	
Foyles, George	Pvt	A			
France, Alex E	Cpl	B		S— Oct. 29-18	Dec. 2-18.
Francis, Henry W	Cpl	I		S— Sept. 6-18	Nov. 26-18.
Franz, George	Pvt	I		W— Sept. 29-18	
Franks, Ersel W	Pvt	MG		S— Nov. 8-18 T— Mar 3-19	Nov. 27-18.
Frazelle, James J	Pvt	B			
Frazelle, Roscoe D	Pvt	H			
Frazier, Martin E	Pvt	B			
Frazier, Roe D	Pvt	B			
Frazier, Van	Pvt--1st Cl	H			
Frederking, Gustav H	Pvt	A		K— Sept. 29-18	
Frederico, Angelo	Pvt	G	Dec. 6-18		
Frederick, John R	Cpl	E		W— Sept. 29-18	Oct. 27-18.
Freeman, Luther M	Pvt	MG	Dec. 6-18	S- -Dec. 27-18	Jan. 5-18.
Freeman, Ralph	Pvt	C		W— Oct. 17-18	Dec. 11-18.
Freeman, Earnley H	Pvt	C		W— Aug. 31-18	
French, Henry	Cpl	L		W— Oct. 10-18	Dec. 2-18.
Friedman, Leo	Pvt	MG		W— Oct. 17-18	

Freto, Antonio	Cpl	M			
Frochlich, Harry B	Cpl	M		Died of Disease --Jan. 20 18	
Fry, John B	Pvt	M		K— Oct. 10-18	
Fry, William G	Pvt	Hq		S— Nov. 11-18	
Fryar, James E	Pvt	G		T— Jul 6-18	
Frisbee, Grover C	Cpl	I			
Fritz, Jesse M	Cpl	D			
Froio, Saverio	Pvt	G	Dec. 6-18		
Fry, William L	Cpl	F		W— Sept. 29-18	Dec. 14-18.
Fulford, Herman D	Pvt	L		S— Oct. 14-18	
Fulford, James I	Pvt	E		K— Oct. 10-18	
Fuchs, Henry J	Pvt	H		W— Oct. 10-18	Dec. 21-18.
Fulford, Samuel G	Pvt	C		W— Oct. 17-18	Nov. 26-18.
Fulk, Alpha	Pvt	MG		S— Nov. 14-18	
Fulghum, Herman W	Pvt	MG	Dec. 6-18		
Fulghum, John D	Mus	Hq			
Fulk, John D	Pvt	A			
Fussell, Raymond D	Sgt	H		K— Sept. 29-18	
Fussnocker, Emil W	Pvt	L		S— Oct. 14-18	

Fuller, Aaron	Cook	C			
Fultz, Leonard	Pvt	E		W— Sept. 28-18	Dec. 11-18.
Furmage, Robert I	Cook	L			
Futrel, Julian H	Pvt	H		S— Oct. 14-18	Oct. 27-18.
Futrell, Albert F	Sgt	C			
Gaddy, Richard C	Pvt	K		S— Oct. 29-18	
Gage, Marion W	Pvt--1st Cl	Hq		W— Oct. 17-18	
Gain, Linden L	Pvt	D		M— Oct. 2-18	
Gaines, William C	Pvt	L		W— Aug. 31-18	Aug. 31-18.
Gaddy, Ernest	Cpl	G		W— Oct. 17-18	Nov. 20-18.
Gaddy, Henry W	Pvt	H		S— Oct. 29-18	Dec. 2-18.
Gaetne, Liberto	Pvt	I	Dec. 6-18		
Gainey, Bert A	Pvt--1st Cl	C			
Gainey, David C	Pvt	H		W— Sept. 29-18	Dec. 2-18.
Galensksa, Peter	Pvt	H			
Gallagher, John T	Sup. Sgt	C			
Gallion, Elmer	Pvt	Hq		W— Aug. 25-18	Aug. 25-18.
Gallion, Eugene	Pvt--1st Cl	M		K— Aug. 5-18	
Galloway, Lewis	Pvt--1st Cl	I			

Gandy, Ernest	Pvt	F		W— Aug. 25-18	Nov. 14-18.
Ganey, James R	Pvt--1st Cl	MG			
Gannon, John	Pvt	I		W— Oct. 17-18	
Ganshow, William	Pvt	M	Dec. 6-18		
Gardener, James H	Pvt	L		K— Sept. 29-18	
Gardner, William	Pvt--1st Cl	L		W— Oct. 10-18	
Gardner, William J	Cpl	K		S— Sept. 26-18	
Gardner, John H	Cpl	D		S— Oct. 1-18	Oct. 29-18.
Gardner, William H	Cpl	I			
Garland, Bascom	Pvt	I			
Garland, Clarence E	Pvt	H		W— Sept. 29-18	
Garland, John L	Pvt--1st Cl	F		K— Sept. 29-18	
Garlow, John H	Pvt	I			
Garner, Carl D	M Sgt	B			
Garner, Danile	Pvt--1st Cl	I			
Garner, Cicero	Pvt--1st Cl	A			
Garner, Cortis H	Pvt--1st Cl	H			
Garner, Alfred	Pvt--1st Cl	I			
Garrand, Albert R	Pvt	A		W— Sept. 29-18	Nov. 26-18

Garren, Volley T	Pvt	H		K— Sept. 29-18	
Garrett, Charles	Pvt--1st Cl	F		S— Oct. 26-18!	
Garrett, Jim H	Pvt--1st Cl	B		W— Sept. 29-18	Nov. 5-18.
Garrett, Joe W	Pvt	B			
Garrett, John H	Cpl	I		W— Sept. 29-18	Dec. 2-18.
Garrett, Manuel	Pvt--1st Cl	B			
Garrett, William H	Pvt--1st Cl	Sup			
Garrett, Grover L	Pvt	K			
Garris, Rufus R	Pvt	B		W— Sept. 29-18	
Garrison, Ernest D	Pvt	G	Dec. 6-18		
Garvin, Alvin E	Pvt--1st Cl	B			
Garvin, Archie C	Pvt	Sup		T— Feb. 26-19	
Gaskins, Sidney E	Pvt--1st Cl	Hq		W— Oct. 17-18	
Gaskins, William O	Pvt	D		W— Sept. 27-18	
Gasparatte, Tony	Pvt--1st Cl	B			
Gathings, James R	Pvt--1st Cl	E			
Gattis, Carver M	Pvt	C			
Gay, Leon S	Cpl	F			
Gaynes, Alfred	Pvt	D			
Gaynus, Neil	Pvt	M			

Geiser, Louis, Jr	Pvt	Hq		W— Oct. 17-18	Dec. 1-18.
Gendreau, Homer J	Pvt	K			
George, Grover C	Pvt	O			
George, Charles W	Ck	A		T— Feb. 4-19	
Giardina, Frank	Pvt--1st Cl	Hq		S— Nov. 11-18	
Gerald, Jennings	Bug	L			
Gereshkofsky, Herny	Pvt	B	Dec. 6-18		
Gerhardt, George	Pvt	D			
Gerlach, August R	Pvt	C		W— Sept. 26-18	Dec. 19-18.
Gher, Louis N	Pvt	Hq			
Gibbons, John F	Pvt	I		W— Sept. 29-18	Dec. 5-18.
Gibbs, Hermend V	Cpl	G			
Gibby, Britt	Pvt	E			
Gibson, Archie W	Sgt	Hq			
Gibson, Roy C	Pvt	I			
Gibson, William E	Sgt	A			
Gibson, Olin E	Cpl	A		K— Oct. 16-18	
Gibson, Sam	Pvt	H		W— Sept. 29-18 S— Mar 15-19	Dec. 19-18.

Giles, Scott E	Sgt	Hq			
Gill, Harley	Pvt	K		W— Sept. 29-18	Oct. 29-18.
Gillahan, Fred A	Sgt	L		K— Sept. 28-18	
Gilliad, William	Pvt	K		K— Oct. 18-18	
Gilliam, Norman L	Pvt	Mg	Dec. 6-18		
Gilliat, Earl E	Pvt--1st Cl	C		W— Aug. 18-18	Sept. 25-18.
Gillikin, Fernie B	Pvt	M			
Gillikin, Thomas L	Pvt--1st Cl	Inf		S— Oct. 27-18	Nov. 26-18.
Gillespie, Claude	Pvt	F		W— Oct. 17-18	Nov. 25-18.
Gillispie, John T	Pvt--1st Cl	M		S— Mar 16-19	
Gilman, Edward	Pvt	F			
Gilmore, John C	Sgt	D		T— Jul 22-18	
Gindling, Joe P	Pvt	I			
Gladden, William H	Pvt	L		W— Oct. 17-18	Dec. 2-18.
Glanton, Mack	Pvt	A			
Glassbrenner, Fred L	Pvt--1st Cl	Hq		K— Sept. 29-18	
Glasscock, Willie T	Pvt	B		S— Oct. 14-18	
Glasscock, John A	Pvt	B		W— Oct. 10-18	Dec. 8-18.
Glenn, Homer O	Pvt	H			

Glenn, John T	Pvt--1st Cl	D			
Glisson, Felix A	Pvt	D			
Glover, James C	Pvt	I		W— Sept. 29-18	
Goble, William R	Pvt--1st Cl	D		W— Oct. 17-18	Dec. 29-18.
Godwin, John R	Cpl	K			
Godwin, Lee	Wag	Sup			
Godwin, Perry	RSS	Sup			
Godwin, Perry L	Sgt	M			
Goeller, Sam W	Pvt	MG	Dec. 6-18		
Goers, Horace C	Pvt	A		S— Oct. 31-18	Dec. 8-18.
Goff, Arza A	Pvt	C		S— Nov. 11-18	Nov. 26-18.
Goggard, Walter S	Sgt	I		K— Sept. 1-18	
Goins, James F	Pvt	G		W— Sept. 29-18	
Goldburg, Sam	Pvt--1st Cl	Hq		W— Sept. 29-18	
Golden, Bennie E	Cpl	B			
Golden, Walter	Pvt	B			
Goldman, Elmer	Pvt	I			
Gooch, William R	Wag	Sup			
Goode, Willie	Pvt--1st Cl	B		W— Oct. 17-18	Dec. 2-18.

Gooden, Charles F	Pvt--1st Cl	MG			
Goodin, Elbert	Pvt	L		W— Oct. 17-18	Dec. 14-18.
Goodman, Albert	Pvt	L		M— Aug. 31-18	
Goodpasture, Elmer L	Pvt	E		W— Sept. 29-18	
Goodson, Wade	Pvt--1st Cl	H		T— Jan. 20-19	
Goodwin, Harry E	Pvt	B		S— Nov. 11-18	Nov. 27-18.
Goodwin, John	Pvt	H	Dec. 6-18	S— Dec. 29-18	Jan. 25-19.
Goodwin, Tallie R	Pvt--1st Cl	MG			
Gordon, Ernest W	Sgt	Sup			
Gordon, James W	Pvt	C		S— Oct. 28-18	Nov. 26-18.
Gordon, Newton B	Pvt	K		S— June 4-18	
Gordon, Tom E	Pvt	E		W— Sept. 29-18	
Gore, James A	Wag	Sup			
Gore, William A	Pvt--1st Cl	Hq		W— Oct. 17-18	Dec. 1-18.
Gorganeous, Floyd D	Sgt	H		W— Sept. 29-18	Oct. 27-18.
Gornelli, Frank	Pvt	H	Dec. 6-18		
Goss, Wilbur R	Pvt	K		W— Oct. 10-18	
Gossman, Henry	Pvt	K		W— Aug. 26-18 K— Oct. 20-18	
Gould, Hiser H	Sgt	A			

Grady, John K	Cpl	B		W— Oct. 17-18	
Grady, Emmet	Pvt	C			
Grady, William O	Pvt--1st Cl	C		W— Sept. 29-18	Dec. 2-18.
Graf, Ernest G	Pvt	Hq		T— Jan. 12-18	
Graff, Charles W	Pvt	Mg		W— Oct. 10-18	
Graham, John W	Sgt	B		T— Sept. 22-18	
Graham, William	Pvt--1st Cl	L		W— Sept. 29-18	
Graham, Reginald G	Pvt--1st Cl	Inf		W & D— Sept. 28-18	
Graham, Daniel F	Cpl	L		W— Sept. 29-18	Dec. 2-18.
Graham, John E	Pvt--1st Cl	MG			
Graham, William E	Pvt	I			
Grainger, Charles S	B Sgt	Hq			
Granade, Carl	Pvt	MG	Dec. 6-18		
Granate, Antonio	Pvt	B	Dec. 6-18		
Grandstaff, Robert F	Wag	Sup			
Graner, Fred	Pvt	E		S— Oct. 29-18	Dec. 3-18.
Grant, Edward P	Pvt	D			
Grant, Lester	Wag	Sup			
Grant, Ernest	Pvt	A		T— Jan. 29-19	

Grantham, Luther L	Pvt--1st Cl	MG		W— Sept. 29-18	Nov. 8-18.
Grantham, Adrian H	Pvt--1st Cl	MG		K— Sept. 29-18	
Gratham, William	Pvt	F		W— Oct. 17-18	
Gratopp, Edward A	Pvt	H	Dec. 6-18		
Graves, Fred T	Sgt	Hq		T— Nov. 16-18	
Gray, Herbert B	Mech	B		W— Oct. 17-18	
Gray, Hilary C	Pvt	F		W— Aug. 25-18	
Grey, Marion J	Pvt	E		S— Nov. 22-18	
Gray, Clarence R	Pvt	MG			
Gray, Elmer	Pvt	MG	Dec. 6-18		
Gray, Robert S	Pvt--1st Cl	G			
Graziani, Angelo	Pvt	H	Dec. 6-18		
Greco, Joe	Pvt--1st Cl	E			
Green, Andy W	Pvt	H		W— Oct. 10-18	Dec. 7-18.
Green, Archie R	Pvt	G			
Green, Brodie L	Sgt	H		W— Sept. 29-18	Oct. 7-18.
Green, Grady	Pvt	G		S— Nov. 22-18	Dec. 23-18.
Green, Jesse H	Cpl	F		W— Sept. 29-18	Oct. 27-18.
Green, Pierson	Cpl	G			

Greene, Jesse C	Wag	Sup			
Green, Daniel H	Pvt--1st Cl	D		W— Oct. 17-18	Nov. 26-18.
Green, Thad	Pvt	G		W— Sept. 29-18 T— Feb. 4-19	
Greene, Milton A	Pvt	A		K— Sept. 29-18	
Green, Walter S	Pvt	H		T— Jul 6-18	
Greer, Edmond D	Cpl	H		K— Sept. 29-18	
Greer, Samuel	Pvt	D		W— Oct. 17-18	
Gregorion, Goust	Pvt	H	Dec. 6-18		
Gregory, James E	Cpl	M		S— Nov. 27-18	Dec. 10-18.
Gregory, Willie T	Pvt	K			
Gries, Fred C	Mech	MG			
Griffin, Gus	Pvt--1st Cl	A		W— Sept 29-18	
Griffion, Charles H	Pvt	H		S— Nov. 11-18	Dec. 21-18.
Griffin, Manuel F	Pvt	M			
Griffin, William C	Pvt	A			
Griffin, William K	Cpl	MG		W— Sept. 29-18	Dec. 10-18.
Grifford, Theodore P	Pvt	G	Dec. 6-18		
Grimes, Earl G	Pvt	F		W— Sept. 29-18	Oct. 29-18.
Grimes, Manly	Pvt	E		W— Sept. 29-18	Dec. 19-18.

Grimes, Noah J	Pvt--1st Cl	A		W— Sept. 29-18	Nov. 11-18.
Griswold, Charles	Pvt	K		W— Oct. 10-18	
Grizzard, Russell	Pvt--1st Cl	MG			
Groce, Maurice E	Pvt--1st Cl	B			
Groening, Frank J	Mech	MG		W— Oct. 10-18	Oct. 27-18.
Groner, William B	Pvt--1st Cl	D		W— Oct. 10-18	
Groskrentz, William E	Pvt	K		K— Oct. 9-18	
Groszewski, Joseph	Pvt	MG	Dec. 6-18		
Grubb, Harmon D	Pvt	K		K— Oct. 8-18	
Grubbs, Ernest G	Pvt	A		W— Oct. 10-18	
Grubbs, Lee	1st Sgt	B			
Guerrettaz, Elsa	Pvt	A		W— Sept. 29-18	
Guns, Thomas H	Pvt	MG	Dec. 6-18		
Gunter, Voice V	Pvt	E		S— Aug. 12-18	Nov. 11-18.
Gurganous, Otway	Pvt	Sup			
Gurganus, Edward P	Sgt	I			
Gurganus, Winston A	Cpl	I		W— Sept. 29-18	Dec. 2-18.
Gurlie, Vernie	Pvt	Sup			
Gussett, Claude	Pvt	K		W— Oct. 10-18	
Gust, William A	Pvt	F		W— Oct. 17-18	Nov. 26-18.

Guth, Fred C	Pvt	Hq		K— Aug. 31-18	
Gutusky, Stanley	Pvt	H	Dec. 6-18		
Guy, Hillie	Pvt	E			
Guyer, Thomas L	Pvt	I		W— Sept. 29-18	Dec. 14-18.
Guyse, Jay G	Pvt	B	Dec. 6-18		
Haag, Fred A	Pvt	F		K— Oct. 10-18	
Haag, Nathan	Pvt	B		W— Oct. 17-18	Dec. 14-18.
Haag, Carl G	Pvt--1st Cl	B			
Hackney, James	Pvt	F			
Hackney, William F	Pvt--1st Cl	F		W— Sept. 29-18	
Haffner, Frank	Pvt	H		W— Sept. 29-18	
Hagan, Luther J	Pvt	K			
Hagans, Gordon S	Pvt	C		T— Mar 10-18	
Hagemann, Edwin E	Pvt--1st Cl	L			
Hagen, Glen A	Cpl	F		M— Oct. 26-18	Dec. 19-18.
Hagler, Charles S	Pvt	M			
Hagler, Daniel C	Cpl	H			
Hagler, Sam E	Pvt--1st Cl	F		W— Oct. 10-18	
Hahn, Leonard G	Pvt	K		S— Oct. 20-18	Dec. 14-18.

Haislet, Aulden	Pvt	K			
Halberg, Edwin O	Pvt	K			
Hale, Clyde A	Sgt	Hq		W— June 20-18	
Hall, Allen J	Pvt	G			
Hall, Ervin	Cpl	L			
Hall, Howard M	Pvt	C			
Hall, James B	Sgt	Hq			
Hall, John	Mus	Hq			
Hall, Lamn H	Pvt	B		W— Oct. 10-18	Dec. 4-18.
Hall, Leon M	Pvt--1st Cl	Hq			
Hall, Leonard J	Pvt	M		S— Nov. 11-18	Dec. 5-18.
Hall, Luther P	Pvt--1st Cl	A		W— Aug. 30-18	Nov. 15-18.
Hall, Seth	Cpl	M			
Hall, Thomas	Pvt	D			
Hall, Will D	Cpl	K		S— Oct. 29-18	Dec. 5-18.
Hall, Banner D	Cpl	F			
Hall, Charlie B	Pvt	G		W & Died Sept. 30-18.	
Hall, Charles E	Pvt	K		K— Sept. 29-18	
Hall, Floyd D	Pvt	H		S— Jul 6-18	

Hall, Henry H	Pvt	L		K— Aug. 31-18	
Hall, Lonnie T	Pvt	E		W— Sept. 29-18	
Hall, Raleigh W	Pvt	D		W— Aug. 21-18	
Hall, William E	Sgt	F		W— Sept. 29-18	
Halterman, Albert L	Pvt	D		K— Oct. 7-18	
Hamby, Verndin H	Pvt	G			
Hamilton, Foster K	Sgt	Hq			
Hamilton, John H	Pvt--1st Cl	G		M— Sept. 29-18	
Hammack, Elbert L	Pvt	L			
Hammock, Andrew B	Pvt--1st Cl	L		S— Sept. 10-18	
Hampton, Oscar	Pvt--1st Cl	Hq		S— Nov. 11-18	
Hampton, Henry W	Pvt	K			
Hampton, James P	Cpl	C			
Hand, Lee	Pvt	MG		W— Oct. 10-18	
Handford, Robert	Pvt	MG			
Hanes, George F	Pvt	MG	Dec. 6-18		
Hankins, Louis M	Pvt	D		W— Sept. 29-18	
Hankins, Marvin J	Cpl	L			
Hannah, John H	Cpl	D		W— Oct. 17-18	

Hannah, Roy H	Pvt	K		T— Jan. 22-19	
Hannel, Jacob R	Pvt	K		K— Oct. 9-18	
Hanningan, Mike	Pvt--1st Cl	L		T— Jul 15-18	
Hansen, John	Pvt	K			
Hanson, Lars	Pvt	H			
Happich, Ferdinand G	Pvt	MG	Dec. 6-18		
Harder, Roy J	Cpl	K			
Harbert, Howard	Pvt	D		W— Sept. 29-18	
Harbin, Marshall C	Pvt	M			
Harbour, Shelby G	Pvt	D		S— Oct. 28-18	
Harden, Valen F	Pvt--1st Cl	I			
Hardin, LaFayette E	Pvt	G		W— Sept. 28-18	Dec. 14-18.
Hardin, Noah L	Pvt	E		K— Aug. 20-18	
Harding, Lawrence B	Pvt	F		W— Oct. 19-18	
Hardison, John M	Pvt--1st Cl	G		W— Oct. 10-18	
Hardison, Asa H	Wag	Sup			
Hardwick, John T	Pvt	G		W— Sept. 29-18	
Hardy, Garley C	Pvt	MG		W- -Sept. 29-18	
Hardy, James F	Pvt--1st Cl	A			

Hardy, Melvin	Pvt--1st Cl	M			
Hardy, William R	Pvt--1st Cl	L		W— Sept. 29-18	Dec. 2-18.
Hardy, William E	Pvt	H		W— Oct. 10-18	
Hare, Edward W	Cpl	Ordn.			
Hager, Orion	Pvt	L		W— Oct. 17-18	
Harges, Cordell	Pvt--1st Cl	B			
Harges, Willie M	Pvt--1st Cl	M		W— Aug. 31-18	Nov. 11-18.
Harkey, Lonnie R	Pvt	Hq	Dec. 6-18		
Harmin, Willie D	Pvt--1st Cl	A		K— Sept. 29-18	
Harmon, James G	Pvt--1st Cl	D			
Harmon, Monroe	Pvt	MG			
Harp, Walter E	Pvt--1st Cl	L			
Harper, Lee E	Ck	F			
Harper, Leon	Pvt--1st Cl	B			
Harrell, Arthur	Pvt	I			
Harrell, Clyde B	Cpl	M			
Harrell, Vance	Pvt	K			
Harrell, Edison W	Pvt	I		S— Aug. 21-18	
Harrell, John H	Cpl	E		W— Sept. 29-18	

Harrell, Roland	Pvt	I		K— Aug. 2-18	
Harris, Dewey S	Cpl	Hq		T— Mar 16-19	
Harris, Ernest G	Pvt	D		S— Oct. 14-18	
Harris, Faison	Mech	D		K— Jul 18-18	
Harris, Loyd T	Pvt	F		W— Set. 29-18	
Harris, Hall C	Pvt--1st Cl	K		W— Oct. 10-18	
Harris, Perley H	Cpl	B		K— Sept. 29-18	
Harris, Victor E	Pvt	C		W— Sept. 18-18	
Harris, William C	Cpl	I		Discharged June 18-18.	
Harris, Barry B	Pvt	E			
Harris, Charlie	Pvt--1st Cl	I		W— Aug. 31-18	Sept. 19-18.
Harris, Frank W	Cpl	L			
Harris, Gaither J	Bug	F		S— Oct. 1.18	Nov. 14-18.
Harris, George W	Pvt--1st Cl	Hq			
Harris, Giles H	Cpl	K		W— Sept. 29-18	Dec. 14-18.
Harris, Gurley W	Cpl	L		S— Oct. 14-18	Nov. 11-18.
Harris, Harry	Pvt	D		W— Aug. 31-18	Dec. 23-18.
Harris, James H	Cpl	E			
Harris, John T	Pvt	D		W— Oct. 17-18	Jan. 18-19.

Harris, Lonnie L	Cpl	B			
Harris, Mack D	Mech	A		W— Oct. 17-18	Nov. 26-18.
Harris, William S	Pvt	E			
Harrelson, Norman J	Pvt	Sup			
Harrenstein, Herman A	Pvt	H	Dec. 6-18		
Harrington, Joe J	Pvt--1st Cl	H		W— Sept. 29-18	Dec. 30-18.
Harrington, William J	Pvt	E			
Harrison, James B	Wag	Sup			
Harrison, Martin P	Sgt	C			
Harrison, Royal A	Sgt	C		T— June 22-18	
Hart, Samuel C	Pvt--1st Cl	G		K— Sept. 29-18	
Hart, Isaac	Pvt	Sup			
Hartig, Ernest H	Pvt	K		W— Sept. 29-18	
Hartley, Donald M	Pvt--1st Cl	MG.			
Hartley, Matt	Pvt	C			
Hartman, Albert	Pvt	H		W— Sept. 29-18	Oct. 10-18.
Hartnels, Edward B	Pvt	M			
Hartsell, Albert A	Pvt	H		M— Sept. 29-18	Nov. 8-18.
Hartsell, Arthur	Pvt	MG.			

Hartshorn, Everet L	Pvt	D		W— Sept. 29-18	
Harvel, Stanley J	Cpl	I			
Harvell, Jacob D	Cpl	A		W— Oct. 10-18	Nov. 26-18.
Harvey, Hiram	Pvt	I		W— Oct 17-18	
Haslach, George F	Pvt	MG.	Dec. 6-18		
Hassell, James L	Pvt	I		W— Sept. 29-18	
Hassler, Lester D	Sup Sgt	G			
Haswell, Odie	Pvt	MG.	Dec. 6-18		
Hatch, James R	Sgt	A		T— Jan. 29-19	
Hatfield, William M	Pvt	H		W— Sept. 29-18	
Hauger, Herbert E	Pvt	K		W— Oct. 10-18	Dec. 2-18.
Hawk, Russell	Pvt	MG.	Dec. 6-18		
Hawkins, Frank	Pvt	M			
Hawkins, Frank R	Pvt--1st Cl	C			
Hawkins, Orville	Pvt--1st Cl	F			
Hawkins, Paul G	Pvt--1st Cl	MG.			
Hawkins, Greeley	Pvt	L		T— Feb. 8-19	
Hawkins, Tandy	Pvt	M		K— Oct. 10-18	
Hawley, Ivey	Pvt	MG			

Hay, Robert C	Wag	Sup			
Haydock, Henry	Pvt	A		K— Oct. 10-18	
Hayes, General	Pvt	C		S— May 1-18	
Hayes, Durward	Pvt	H			
Hayes, Grady S	Wag	Sup			
Hayes, Roland H	Ck	I			
Haymond, Harvey K	Pvt	F		W— Sept. 29-18	
Haynes, James F	Pvt	E		W— Sept. 29-18	
Haynes, Thomas M	Pvt--1st Cl	C		W— Oct. 17-18	
Haynes, Adrien D	Pvt	F			
Haynes, Clarence D	Cpl	Hq			
Haynes, John S	Pvt--1st Cl	C			
Haynes, James A	Pvt	C		W— Sept. 29-18	
Haynes, Leonard	Pvt	D			
Haynes, William H	Pvt	E			
Haynes, Winfred F	Pvt	H			
Hays, Parks L	Sgt	E			
Hays, Thomas H	Pvt--1st Cl	E		S— Nov. 22-18	Dec. 2-18.
Hays, Webb B	Bug	C			

Hazlett, Burt	Pvt--1st Cl	Hq		W— Sept. 29-18	Dec. 1-18.
Head, William B	Pvt	Sup			
Hearlston, Jasper G	Pvt	E		W— Sept. 29-18	
Heath, William H	Pvt--1st Cl	G		W— Jul 18-18	
Heavner, Henry E	Pvt--1st Cl	I			
Hedgepeth, Lee	Pvt	MG.			
Hedjuk, Thomas	Pvt	B			
Heath, Elder E	Cpl	I		W— Sept. 29-18	Dec. 2-18.
Hedrick, Neal A	Pvt	G			
Hefinger, George A	Cpl	C		T— Feb. 14-19	
Heflin, William G	Pvt	D		W— Sept. 1-18 D— Sept. 2-18.	
Heflin, Aubrey	Pvt	E			
Hein, Charles F	Pvt	H		W— Oct. 10-18	Oct. 10-18.
Hellman, William J	Pvt	K		S— Oct. 14-18	Nov. 11-18.
Helms, Tom	Pvt	H		M— Sept. 29-18	Dec. 2-18.
Helton, Benjamin	Wag	Sup			
Helton, Carl H	Cpl	Hq		S— Oct. 29-18	Nov. 11-18.
Hemann, Herman H	Sgt	F			
Hemmerick, Edmund L	Pvt	A	Dec. 6-18		

Henderson, Charlie	Pvt	M		S— Nov. 11-18	Dec. 3-18.
Henderson, Ernest J	Pvt	K			
Henderson, Willie H	Ck	Hq			
Henderson, Grover C	Ck	MG.			
Henderson, Harley	Pvt	M		K— Oct. 10-18	
Henderson, Herbert W	Sgt	F		W— Oct. 10-18	
Henderson, Luby L	Pvt	H		S— Jul 6-18	
Hendley, Nimrod	Cpl	Hq		W— Sept. 29-18	Sept. 29-18.
Hendricks, Isaac D	Cpl	A		W— Sept. 29-18	Dec. 14-18.
Hendricks, Samuel G	Pvt--1st Cl	MG.		W— Aug. 25-18 T— Feb. 12-19	
Hendrix, Edwards H	Pvt	F		W- Sept. 29-18 S— Oct. 14-18	
Hendrix, Herman	Pvt	F		W— Oct. 10-18	
Hendron, Charles C	Pvt--1st Cl	H		W— Sept. 29-18	Nov. 11-18.
Henries, James M	Pvt--1st Cl	L			
Henry, Barney A	Cpl	C		W— Oct. 10-18	Nov. 20-18.
Henry, Robert E	Sgt	Hq			
Henry, Roy O	Pvt--1st Cl	Hq			
Henry, Herbert H	Cpl	L		K— Oct. 14-18	

Henry, Oakey H	Pvt	D		W— Oct. 10-18	
Hensley, Mack	Pvt	I		W— Aug. 31-18	Dec. 2-18.
Herman, Kenneth S	Sgt	D		W— Oct. 10-18	Dec. 2-18.
Hermoach, Alvin L	Cpl	Sup			
Herrin, William H	Cpl	Hq			
Herring, Fred	Ck	H			
Herrman, Louis	Pvt	C		W— Sept. 3-18	Oct. 1-18.
Herthel, George W	Cpl	L			
Hesson, Oakey E	Pvt	D		W— Aug. 31-18	
Hewitt, William G	Sgt	Hq		K— Aug. 31-18	
Hewitt, William T	Pvt	B		W— Sept. 26-18 T— Jan. 22-19	Dec. 11-18.
Hibbard, Albert L	Sgt	K		W— Sept. 29-18	Nov. 20-18.
Hickman, Henry C	Sgt	G		T— Jul 22-18	
Hickman, Julius L	Cpl	G		Sept. 29-18	
Hickman, William T	Pvt	Inf		K— Aug. 31-18	
Hicks, Grant C	Pvt	B		T— Feb. 4-19	
Hicks, John	Pvt	A		W— Sept. 29-18	
Hicks, Robert L	Pvt	D		W— Sept. 29-18	
Hicks, William	Pvt	A		K— Oct. 10-18	

Hicks, Burt	Cpl	Hq			
Higgins, Hiram M	Cpl	MG.		W— Sept. 27-18	Oct. 27-18.
Hilburn, Norman	Pvt	B			
Hildreth, Arpie M	Wag	Sup			
Hileman, John N	Pvt	E		W— Sept. 29-18	
Hill, Buck	Pvt	L		K— Sept. 29-18	
Hill, Charlie W	Pvt	D		W— Oct. 17-18	
Hill, John H	Pvt	Inf		S— Oct. 3-18	
Hill, Rowden W	Pvt	L		W— Sept. 29-18	
Hill, Clarence O	Pvt	MG.			
Hill, Elzie	Pvt	I			
Hill, Louis H	Cpl. Sgt	Hq			
Hill, Simmie H. D	Pvt--1st Cl	A		W— Oct. 10-18	Dec. 10-18.
Hill, William W	Pvt	A			
Hilles, Adolph E	Pvt	K			
Hillig, Harry	Pvt	G		K— Sept. 29-18	
Hinds, Dorsey W	Pvt--1st Cl	Hq			
Hinds, Estill	Sgt	B			
Hinds, James	Pvt	B		W— Oct. 17-18	Nov. 11-18.

Hinds, Roscoe C	Sgt	Hq			
Hinds, Taylor	Wag	Hq		W— Oct. 17-18	Nov. 11-18.
Hines, Robert W	Cpl	F			
Hinson, James W	Sgt	D			
Hinson, Noah A. L	Pvt	I		S— Nov. 22-18	Dec. 11-18.
Hinson, Sindey W	Cpl	D		K— Oct. 10-18	
Hinton, Jake	Pvt	M		S— Jul 1-18	
Hinton, Orville W	Pvt--1st Cl	Hq		S— Nov. 11-18	Nov. 26-18.
Hipp, John L	Pvt	MG.			
Hippard, David G	Mech	Hq			
Hoag, Edwin F	Pvt--1st Cl	L			
Hobba, Robert E	Pvt	E			
Hobbs, Benjamin S	Ck	G		W— Sept. 29-18 T— Jan. 29-19	Nov. 11-18.
Hockett, James L	Pvt	MG.			
Hocking, Ernest	Pvt	L		W— Oct. 10-18	
Hodge, Sam J	Pvt	K			
Hoffman, Hurt	Pvt	F			
Hoffman, Raymond H	Pvt	K		W— Sept. 29-18	
Hofler, Charlie W	Pvt--1st Cl	D		K— Oct. 22-18	

Hogan, Harry L	Ck	B			
Hogan, Walter	Pvt--1st Cl	Hq			
Hoggard, Joe W	Cpl	K		K— Oct. 6-18	
Hoglan, Everett B	Pvt	F		W— Oct. 10-18	Dec. 2-18.
Hogsed, Paul	Pvt	I		K— Aug. 2-18	
Holbrook, John L	Cpl	F		M— Oct. 10-18	Oct. 29-18.
Holcombe, James R	Pvt	I		W— Aug. 31-18	Nov. 11-18.
Holcombe, Reginald M	Pvt	H			
Holder, Allen G	Pvt	H	Dec. 6-18		
Holder, Marvin K	Wag	Sup			
Holder, Albert G	Pvt	C		W— Jul 18-18	
Holder, Edward B	Pvt	G		W— Sept. 29-18	
Holder, John F	Pvt--1st Cl	C		W— Sept. 29-18	
Holland, George R	Wag	Sup		S— Dec. 8-18	Dec. 10-18.
Holland, John	Pvt	A	Dec. 6-18		
Holland, Miles	Pvt	A		S— Jul 6-18	Oct. 27-18.
Holland, Robert F	Mus	Hq			
Holleman, Frank P	Pvt--1st Cl	F		T— Sept. 12-18	
Hollingsworth, Alfred	Mess Sgt	F		S— Mar 16-19	

Hollingsworth, John P	Cpl	L		S— Nov 1-18	
Hollingsworth, Luby G	Cpl	Hq			
Hollingsworth, Luther O	Mus	Hq			
Hollmann, Henry T	Pvt	D		S— Mar 15-19	
Holloman, Albert O	Pvt	K			
Hollon, Owen	Pvt--1st Cl	L		W— Oct. 17-18	
Holloway, Christopher C	Pvt	Hq			
Holmes, Joseph	Pvt	A		S— Oct. 29-18	
Holmes, Albert C	Pvt	I			
Holmes, Ed	Pvt	A		W— Sept. 29-18	
Holtzman, Isadore	Pvt	E		W— Sept. 29-18	Dec. 2-18.
Holyfield, Early L	Pvt--1st Cl	D		W— Oct. 10-18	
Honeycutt, Grover C	Pvt	MG		S— Dec. 8-18	
Honeycutt, Lucian G	Pvt	H		S— Oct. 31-18	
Honeycutt, Ralph	Pvt	A		W— Sept. 29-18	
Honeycutt, Benjamin	Pvt	Sup			
Honeycutt, Eli	Pvt	M			
Honeycutt, Frank W	Pvt	Hq			
Honeycutt, Reddin E	Cpl	M		S— Oct. 31-18	Nov. 26-18.

Honeycutt, William A	Wag	Sup			
Honnecker, Harmin L	Cpl	Hq		S— Nov. 11-18	Nov. 26-18.
Hood, William L	Cpl	D		W— Sept. 29-18 S— Mar 16-19	Nov. 12-18.
Hood, Henry C	Pvt	B			
Hood, Howard A	Cpl	C			
Hooper, Monroe C	Pvt	I			
Hooper, Jesse	Cpl	E		S— Oct. 16-18	Dec. 2-18.
Hooper, Joe	Pvt--1st Cl	I		S— Oct. 14-18	Nov. 1-18.
Hope, Henry G	Pvt	Hq	Dec. 6-18		
Hopkins, Frank	Pvt	E		K— Oct. 10-18	
Hopkins, Lonnie D	Cpl	L		W— Sept. 29-18	Dec. 11-18.
Horn, Carroll W	Pvt	K		W— Sept. 29-18	
Horn, Erasmus	Pvt	G		S— Oct. 28-18	
Horn, Grod C	Ck	Sup			
Horn, Ralph	Pvt	Hq			
Horne, Irvery M	Pvt--1st Cl	A		W— Oct. 17-18	
Horne, Joe R	Pvt	D		W— Oct. 17-18	
Horne, Edward B	Cpl	Hq			
Horne, Thomas L	Pvt	M			

Horsley, Iran E	Pvt	I		S— Oct. 31-18	
Hoskins, Baker	Pvt	F			
Honck, Charley O	Pvt	M			
Houghton, Elias B	Pvt	D		W— Oct. 10-18	
Houk, James L	Pvt	I			
Houser, Boulah	Pvt--1st Cl	L		T— Jan. 29-19	
Houser, Joshua P	Pvt--1st Cl	G		W— Jul 29-18 T— Jan. 29-19	Dec. 2-18.
Houston, Thomas W	Wag	Sup		S— Mar 16-19	
Houston, James N	Pvt	E		W— Sept. 28-18	Sept. 28-18.
Howard, Harvey G	Pvt	B		S— Jul 6-18	
Howard, Phillip	Pvt	G		W— Sept. 29-18	
Howard, Arch	Pvt--1st Cl	Hq		W— Sept. 4-18	Dec. 1-18.
Howard, Charles H	Pvt	E			
Howard, Fred	Pvt	G			
Howard, John T	Cpl	I			
Howard, William L	Pvt	H	Dec. 6-18		
Howarth, James	Pvt	C		W— Aug. 31-18	Dec. 4-18.
Howell, Eddie	Pvt	A		W— Oct. 17-18 T— Feb. 8-19	

Howell, Francis M., Jr	Pvt--1st Cl	E		W— Oct. 10-18	
Howell, Jay D	Pvt	O		W— Sept. 1-18 W— Oct. 10-18	
Howell, John I	Sup. Sgt	A		T— Jan. 29-19	
Howell, Maury	Mech	B		S— Oct. 28-18	
Howell, Lonnie H	Sgt	E			
Howell, McKinley R	Pvt	G			
Howell, Willie F	Pvt	A			
Howerton, George E	Pvt	F			
Howle, Robert N	Pvt--1st Cl	Inf		S— Sept. 11-18	Sept. 15-18.
Howle, Edwin P	Pvt--1st Cl	Inf		W— Sept. 29-18	
Hubbard, Albert L	Pvt	I			
Hubbard, Sidney	Mech	F			
Hubbard, Mack	Cpl	A		W— Sept. 29-18	
Hubert, Anthony M	Pvt	MG		W— Sept. 29-18	
Huddleston, Oren W	Pvt--1st Cl	G		T— Jan. 22-19	
Hudson, Alma H	Pvt	A		W— Oct. 17-18	Nov. 26-18.
Hudson, Ernest B	Pvt	A		W— Sept. 29-18	
Hudson, Arthur E	Pvt	H	Dec. 6-18		
Hudson, Frank B	Pvt	L			

Hudson, Gillard H	Ck	H			
Hudson, Park	Ck	M			
Hudson, Joe	Pvt	H		S— Sept. 6-18	
Hudson, Norman	1st Sgt	B		T— Jul 22-18	
Huffstetler, Oscar E	Pvt	H		W— Sept. 29-18	Nov. 11-18.
Hufham, Morris V	Sgt	E		T— Jan. 22-19	
Huggins, Walter G	Cpl	H			
Hughes, Claude A	Cpl	M			
Hughes, Clifford C	Pvt	C		S— Oct. 28-18	Dec. 21-18.
Hughes, Dorsey	Pvt	F			
Hughes, John M	Mus	Hq			
Hughes, Julian T	Sgt	MG			
Hughes, Ora C	Pvt	G		W— Sept. 29-18	Nov. 26-18.
Hughes, Patrick F	Pvt--1st Cl	Hq		W— Oct. 17-18	Dec. 1-18.
Hughes, Will	Pvt	H		S— Jul 6-18	Dec. 11-18.
Hughes, Rush H	Sgt	E		W— Oct. 10-18	
Hummell, Samuel G	Sgt	MG		W— Sept. 29-18	
Hummel, Rufus J	Cpl	Hq	Dec. 6-18		
Hudspeth, Walter F	Cpl	E		W— Sept. 29-18	Dec. 14-18.

Hull, Carl L	Pvt	L		W— Sept. 29-18	Oct. 29-18.
Hull, Wade	Pvt	D		S— Oct. 28-18	Dec. 2-18.
Hull, William S	Cpl	H		S— Nov. 22-18	Dec. 2-18.
Hunt, Charles R	Pvt	I			
Hunt, Cicero H	Sgt	Hq			
Hunt, Mose E	Pvt--1st Cl	C		W— Aug. 31-18	Dec. 2-18.
Hunt, Lonnie	Pvt	A		W— Oct. 17-18	
Hunte, George H	Pvt	A		W— Sept. 29-18	Dec. 14-18.
Hunter, Ernest H	Pvt	M			
Hunter, John	Pvt	L		S— Nov. 14-18	Nov. 16-18.
Hunter, Robert W	Pvt	A			
Hunter, Willie	Pvt	E		S— Jul 6-18	
Hurley, Preston	Pvt	H		S— Nov. 11-18	Dec. 7-18.
Hurst, Henry T	Pvt	C			
Hurst, Thomas T	Pvt--1st Cl	G		W— Sept. 29-18	Oct. 17-18.
Hurst, S. Zeno	Pvt	L		W— Oct. 17-18	
Hurt, William L., Jr	Pvt	K			
Hurthel, George	Pvt	L			
Huseth, Erick	Mech	Hq			

Hutcheson, Forrest B	Pvt	D		S— Oct. 1-18	
Hutson, Iven	Pvt	D		W— Sept. 29-18	Dec. 7-18.
Hutson, Edward	Cpl	L		S— Oct. 30-18	
Hyman, Luther D	Pvt	C			
Hyman, Jesse C	Stb. Sgt	C		W— Oct. 17-18	
Hyneman, Urveal	Pvt	L		W— Oct. 17-18	
Hytry, Peter J	Cpl	B		T— Feb. 4-19	
Ingram, William C	Pvt--1st Cl	M		K— Oct. 10-18	
Ipock, Bryce B	Pvt	D		W— Jul 21-18	
Ionora, Nicholas	Pvt	MG	Dec. 6-18		
Inboden, Arthur	Pvt	H	Dec. 6-18		
Ingold, Torrence	Corp	A			
Ingram, Archie	Sgt	I			
Ingram, John H	Cpl	MG			
Ingram, John J	Pvt	Hq		S— Nov. 11-18	Dec. 21-18.
Inman, Perry O	Pvt--1st Cl	MG		S— Sept. 20-18	Dec. 7-18.
Ipock, Reby J	Pvt	D			
Ireland, Clayton J	Pvt	H	Dec. 6-18		
Isbell, Frank T	Pvt	H		S— Oct. 29-18	Dec. 21-18.

Iseley, Alfred H	Pvt	B			
Ives, Charlie W	Pvt--1st Cl	E		S— Oct. 29-18	Dec. 14-18.
Izlar, James F	Pvt	MD			
Jackson, Charles E	Pvt	G		W— Sept. 29-18	Nov. 11-18.
Jackson, Doctor T	Pvt	I			
Jackson, James O	Cpl	M			
Jackson, Luther	Pvt	I		W— Sept. 29-18	Nov. 20-18.
Jackson, Wallace F	Pvt	G		W— Oct. 17-18	Nov. 5-18.
Jackson, Clarence B	Pvt	K		W— Sept. 29-18	
Jackson, Donnie A	Pvt	H		S— Oct. 14-18	
Jackson, Fred T	Pvt	G		K— Sept. 29-18	
Jackson, Jasper D	Cpl	M		T— Feb. 5-19	
Jackson, Lewis	Pvt	E		W— Oct. 10-18	
Jahns, Ernest A	Pvt	B			
Jacobs, William A	Pvt	MD	Oct. 28-18		
James, Jesse	Pvt	H		W— Oct. 10-18	Dec. 2-18.
James, Jesse R	Sgt	D		S— Oct. 14-18	Dec. 14-18.
James, Mark M	Pvt	K			
Jamison, Harry	Pvt	G		W— Sept. 28-18	Dec. 2-18.

Janzeyk, Felix	Pvt	H	Dec. 6-18		
Janowuski, John S	Cpl	A			
Jarrett, Clyde T	Pvt--1st Cl	B			
Jared, George C	Pvt	E		W— Oct. 10-18	
Jay, Joseph	Pvt	L		W— Oct. 17-18	Oct. 17-18.
Jeglie, Walter C	Pvt	K		S— Oct. 31-18	Dec. 17-18.
Jenkins, Harry C	Pvt	Hq			
Jenkins, James L	Pvt	M			
Jenkins, Emmett J	Pvt	H		K— Sept. 29-18	
Jenkins, Van	Ck	Sup		S— Mar 16-19	
Jennerson, George L	Pvt	MG		W— Oct. 29-18	
Jensen, Conrad G	Pvt	F		W— Sept. 29-18	
Jennings, Thomas E	Pvt	B			
Jennings, Truman R	Pvt	C		W— Aug. 31-18 W— Jul 22-18	Sept. 5-18.
Jensen, Ernest E	Cpl	B			
Jensen, Hans	Pvt	MG		W— Sept. 29-18	Nov. 14-18.
Jernigan, Osie C	Pvt--1st Cl	M			
Jethro, James E	Pvt	K		K— Sept. 29-18	
Jeske, Junius J	Pvt	B	Dec. 6-18		

Jemme, James O	Pvt	M		W— Oct. 10-18	Dec. 8-18.
Jett, Thomas B	Pvt	E		S— Nov. 11-18	Dec. 19-18.
Jewell, Joseph W	Pvt	E			
Jewell, Avery	Pvt	H		W— Sept. 29-18	
Jillson, John W	Pvt	L		K— Sept. 29-18	
Jinks, Walter D	Pvt--1st Cl	MD			
Jobe, Chester S	Pvt	Hq		W— Sept. 29-18	
Johns, Albert L	Pvt	H	Nov. 11-18		
Johnson, Alexander L	Pvt--1st Cl	L			
Johnson, Arthur	Pvt	H	Dec. 6-18		
Johnson, Carl E	Pvt	H	Dec. 6-18		
Johnson, Daniel S	Pvt	D			
Johnson, Dewey	Pvt	K			
Johnson, George E	Pvt--1st Cl	B		S— Oct. 1-18	Oct. 7-18.
Johnson, George F	Pvt--1st Cl	L			
Johnson, George P	Cpl	F			
Johnson, Henry	Pvt--1st Cl	D		S— Nov. 22-18	Nov. 25-18.
Johnson, Ira B	Pvt	G		S— Oct. 29-18	Nov. 20-18.
Johnson, James A	Pvt--1st Cl	C			

Johnson, Julius V	1st Sgt	M			
Johnson, Lee	Pvt	F	Oct. 26-18		
Johnson, Lowell W	Pvt	MD	Oct. 28-18		
Johnson, Melvin H	Pvt	H	Dec. 6-18		
Johnson, Oscar E	Pvt	C		W— Sept. 26-18	Dec. 14-18.
Johnson, Roy R	Sgt	C			
Johnson, Thomas C	Pvt	C		S— Oct. 28-18	Dec. 7-18.
Johnson, Wade H	Pvt	H		W— Aug. 26-18	Oct. 29-18.
Johnson, Alvin	Pvt	MG		K— Sept. 29-18	
Johnson, Elder E	Pvt	K		T— Jul 19-18	
Johnson, Ervin M	Cpl	M		W— Aug. 30-18 W— Aug. 31-18	
Johnson, Harry A	1st Sgt	F		T— Jul 22-18	
Johnson, James C	Pvt--1st Cl	L		T— Mar 15-19	
Johnson, Joe	Pvt--1st Cl	E		W— Aug. 31-18	
Johnson, Junius	Cpl	O		W— Aug. 31-18	
Johnson, Carl D	Pvt	D		W— Oct. 17-18	
Johnson, Marcus W	Pvt	A		T— Jul 6-18	
Johnson, Peter	Pvt	H		T— Jul 12-18	
Johnson, Richard	Pvt	H		K— Oct. 10-18	

Johnson, Theodore T	Pvt	D		S— Feb. 21-18	
Johnson, Julius C	Cpl	Hq		W— Sept. 29-18	Oct. 27-18.
Johnston, William B	Pvt	I		K— Aug. 2-18	
Johnston, Arthur C	Pvt	E		W— Sept. 29-18	
Johnston, Matthew E	Cpl	A		W— Sept. 29-18 T— Feb. 4-19	
Jones, Albert F	Pvt	MG		S— Dec. 2-18	
Jones, Andrew J., Jr	Pvt	A		S— Oct. 14-18	
Jones, Clinton V	Pvt	L		S— Jan. 1-19	
Jones, David B	Pvt	E		M— Oct. 17-18	
Jones, David F	Pvt	A		S— Oct. 29-18	
Jones, Fred O	Pvt	C		W— Oct. 17-18	
Jones, Hosey G	Pvt--1st Cl	B		W— Oct. 17-18 T— Feb. 4-19	
Jones, Ira	Pvt	F		K— Oct. 10-18	
Jones, Paul T	Sgt	A		W— Oct. 29-18	
Jones, Waverly B	Opl	L		S— Oct. 30-18	
Jones, Wiley D	Pvt	A		S— Oct. 14-18	
Jones, Willie L	Pvt	F		W— Sept. 29-18	
Jones, Alex	Pvt--1st Cl	B		W— Sept. 29-18	Nov. 11-18.

Jones, Charles M	Pvt--1st Cl	K		S— Oct. 31-18	Nov. 11-18.
Jones, Charlie G	Cpl	B		W— Oct. 10-18	Nov. 27-18.
Jones, Claude	Pvt	E			
Jones, Clifton C	Pvt	A			
Jones, Clyde T	Pvt	Hq			
Jones, Edward	Cpl	D			
Jones, Elmer	Pvt	Hq			
Jones, Fred E	Pvt	A			
Jones, Fred J	Sgt	A			
Jones, Greely D	Cpl	C			
Jones, Harry M	Cpl	H			
Jones, James T	Pvt	A	Dec. 6-18		
Jones, James W	Pvt--1st Cl	I		S— Sept. 13-18	Oct. 27-18.
Jones, Jesse B	Cpl	E			
Jones, Jesse L	Pvt	L		W— Sept. 29-18	Dec. 14-18.
Jones, John W	Wag	Sup			
Jones, Joseph E	Pvt	K		S— Nov. 29-18	Dec. 10-18.
Jones, Maley	Sgt	M		T— Oct. 1-18	Jan. 8-19.
Jones, Paul M	Pvt	K		W— Sept. 29-18	Nov. 20-18.

Jones, Ralph J	Pvt	MG		W— Sept. 29-18	Nov. 26-18.
Jones, Richard N	Pvt	B			
Jones, Roscoe	Pvt--1st Cl	M			
Jones, Roscoe	Pvt	L			
Jones, Selby S	Bd. Ldr	Hq			
Jones, William B	Pvt	L			
Jones, William C	Pvt	F			
Jones, William G	Cpl	C			
Jonican, Owen J	Pvt	D			
Jordan, Auty	Cpl	B			
Jordan, Ben W	Pvt--1st Cl	Hq			
Jordan, Claude P	Cpl	I		S— Oct. 14-18	Nov. 11-18.
Jordan, Robert E	Cpl	E			
Jordan, Charles F	Pvt	H		K— Sept. 29-18	
Jordan, Earl	Pvt	E		S— Nov. 11-18	
Jordan, William A	Pvt--1st Cl	F		W— Sept. 29-18	
Jordan, William C	Pvt	D		W— Jul 20-18	
Joseph, Raymond F	Pvt	Hq		T— Jan. 20-18	
Jorgenson, Fred M	Pvt	H	Dec. 6-18		

Justice, Roy A	Pvt--1st Cl	A		S— Nov. 11-18	
Joseph, Harold J	Pvt	F	Oct. 26-18		
Joseph, Walter V	Cpl	B		W— Aug. 20-18 W— Oct. 10-18	Nov. 26-18.
Joseph, William H	Pvt--1st Cl	Hq			
Josey, Herman O	Pvt	H			
Joyner, Clyde	Pvt	MG			
Joyner, Eddie W	Pvt	K			
Joyner, Gilbert H	Cpl	E		S— Oct. 31-18	Nov. 14-18.
Judd, Joe	Ck	B			
Junge, Peter N	Pvt--1st Cl	C		W— Sept. 29-18	Nov. 14-18.
Juzynski, Leon	Pvt	B	Dec. 6-18		
Kehnke, Sebastin L	Pvt	C		T— Jul 19-18	
Kales, Peter D	Pvt	H		W— Aug. 20-18	Nov. 11-18.
Kamienski, John	Pvt	H	Dec. 5-18		
Kammerer, William E	Pvt--1st Cl	D		W— Oct. 10-18	Nov. 26-18.
Kans, Roy T	Pvt	K		K— Oct. 14-18	
Kanniard, Charles M	Cpl	B		T— Jan. 29-19	
Karbitoff, Charles J	Pvt	L			
Kay, Allen F	Pvt	Hq	Dec. 6-18		

Keavney, Patrick	Pvt	K		W— Oct. 10-18	Dec. 5-18.
Keel, Owen H	Pvt--1st Cl	L		W— Sept. 29-18	
Keen, Houston	Pvt	K		K— Sept. 29-18	
Keen, John O	Pvt	Sup			
Keen, Oker	Cpl	M		W— Oct. 10-18	
Keenan, John R	Pvt	D		K— Sept. 29-18	
Keewosh, John G	Pvt	G		S— Oct. 1-18	
Keiger, Hillary C	Cook	F			
Keill, Ornton A	Pvt	I	Dec. 6-18		
Keller, Frederick	Pvt	A	Dec. 6-18		
Keller, Martin W	Pvt--1st Cl	MG			
Kelley, Edward N	Pvt	I		W— Sept. 29-18	
Kelley, Henry	Pvt	G		W— Oct. 10--18	Dec. 8-18.
Kelly, Carl	Hq	Sup			
Kelly, Charles M	Sgt	F			
Kelly, Charles R	Cpl	H		W— Jul 29-18	
Kelly, Dorsey C	Pvt--1st Cl	Hq		W— Aug. 30-18	Dec. 1-18.
Kelly, Duncan A	Pvt	M			
Kelly, Herbert	Pvt--1st Cl	E			

Kelly, Noah H	Pvt--1st Cl	A		W— Oct. 17-18	Dec. 14-18.
Kendrick, Paul T	Pvt	A	Dec. 6-18	T— Feb. 5-19	
Kennedy, Albert	Pvt	L		W— Aug. 31-18	
Kennedy, Chesley R	Pvt	H		W— Sept. 29-18	Dec. 14-18.
Kennedy, Jack P	Pvt	B	Dec. 6-18		
Kennedy, John L	Sgt	B		T— Jan. 22-19	
Kennedy, Lewis	Pvt	G		W— Oct. 17-18	Dec. 7-18.
Kennedy, Ray R	Pvt--1st Cl	C		W— Sept. 25-18. W— Oct. 17-18	
Kennedy, Walter H	Pvt--1st Cl	B		W— Sept. 29-18	
Kennedy, William	Pvt--1st Cl	B			
Kennerly, Harold L	Pvt	MD			
Kenney, Pat	Pvt	K		S— Nov. 29-18	Dec. 2-18.
Kepler, Lee	Pvt	F			
Kerenz, Thomas R	Pvt	L		S— Nov. 11-18	Feb. 5-19.
Kern, Julian O	Cpl	C		W— Aug. 31-18	Sept. 7-18.
Kernitsky, Sirgey	Pvt	B	Dec. 6-18		
Kersey, Willie	Pvt	E		W— Sept. 29-18	
Kessenger, Clinton R	Pvt	G		W— Oct. 18-18	
Keylon, Robert E	Pvt	E		W— Sept. 29-18	Dec. 11-18.

Keznjin, Thomas	Pvt	H	Dec. 6-18		
Kilbane, Michael	Pvt	K		W— Sept. 29-18	Dec. 5-18.
Kilby, James G	Cpl	M		S— Mar 16-18	
Killian, Sidney E	Pvt--1st Cl	Hq			
Kilpatrick, George L	Pvt	B		S— Mar 16-18	
Kimball, James F	Pvt	K			
Kimball, Joe W	Pvt	E		S— Sept. 29-18	
Kincaide, James W	Cpl	A		W— Oct. 17-18	
Kinery, Brock M	Cpl	I		W— Sept. 29-18	Dec. 5-18.
Kimrey, Otto	Pvt--1st Cl	I			
Kimes, Clarence F	Sgt	K		W— Sept. 29-18	Dec. 5-18.
Kindlesparker, Bert	Pvt	A	Dec. 6-18		
Kindley, John A	Pvt	L		W— Oct. 10-18	
Kindley, Walter D	Pvt	C		W— Aug. 31-18	Aug. 31-18.
Kindrick, William D	Pvt	Hq			
King, Arthur	Pvt	Hq			
King, Bennie F	Cpl	G		W— Sept. 29-18	Nov. 1-18.
King, Ernest H	Pvt	A		K— Oct. 10-18	
King, Everett F	Pvt	B		T— June 1-18	

King, Guy	Pvt--1st Cl	F		W— Oct. 17-18	
King, Hugh H	Pvt	MD			
King, James R	Sgt	M		W— Sept. 29-18	
King, Jasper R	Pvt	C		W— Sept. 29-18	
King, John F	Pvt	H	Dec. 6-18		
King, Joseph D	Pvt	C			
King, Oscar J	Pvt	A		W— Oct. 17-18	
King, Rufus A	Pvt	F		T— May 1-18	
King, Wilbur	Pvt	B			
King, William J	Pvt	G			
Kingston, John H	Pvt	D		W— Sept. 29-18	
Kingston, Raymond J	Pvt	Hq			
Kington, Albert B	Mech	G		W— Sept. 29-18	Dec. 2-18.
Kington, John M	Sgt	G		S— Nov. 11-18	Jan. 15-18.
Kenney, John H	Pvt--1st Cl	Hq		W— Sept. 29-18	Nov. 11-18.
Kenney, Lindo F	Pvt--1st Cl	C		K— Aug. 31-18	
Kinsler, Charles U	Pvt	MD			
Kipley, James H	Pvt	D		W— Sept. 3-18	Dec. 2-18.
Kirkland, Jacob F	Pvt	M		W— Sept. 29-18	

Kirkland, Thomas R	Pvt--1st Cl	Hq			
Kite, Robert K	Pvt	L		W— Sept. 29-18	Dec. 2-18.
Kizziaa, James A	Pvt--1st Cl	Hq		W— Sept. 2-18 W— Sept. 29-18	Nov. 1-18.
Klabunder, William	Pvt	F		W— Aug. 20-18	Dec. 14-18.
Kleebuger, Arthur R	Pvt	M			
Kleffner, Warner H	Pvt	H		W— Sept. 29-18	
Klemme, Valentine	Cpl	L		W— Oct. 17-18	Nov. 11-18.
Kleonpkin, William H	Pvt	M		W— Oct. 10-18	
Klink, John	Pvt	M	Oct. 28-18		
Klucas, Arthur E	Pvt	K		W— Sept. 29-18	
Klutts, Henry C	Pvt	C		W— Oct. 17-18	
Knau, Harley G	Pvt	MD	Oct. 28-18		
Knight, Earl	Cpl	MG			
Knight, Martin	Pvt	H	Dec. 6-18		
Knight, Wallace B	Hs	Hq			
Knight, William L	Pvt	A			
Knight, Willie	Pvt	L			
Knopp, Lewis C	Cpl	D		W— Oct. 17-18	

Knowlton, Rex R	Pvt	Hq	Nov. 4-18	S— Jan. 12-19	
Knowlton, Samuel R	Pvt	A	Dec. 6-18		
Kodix, Mike	Pvt	H	Dec. 6-18		
Koehn, Walter E	Pvt--1st Cl	C		W— Aug. 21-18 W— Sept. 29-18	
Knoelling, Albert F	Pvt--1st Cl	Hq			
Koger, Charles W	Pvt	D		W— Sept. 29-18	
Kohlenberg, Milton P	Pvt	A	Dec. 6-18		
Kohm, William	Pvt--1st Cl	L			
Kometes, John	Pvt	B		W— Oct. 17-18	
Kolaczeski, Stefan	Pvt	H	Dec. 6-18		
Kooistra, Clarence	Pvt	H	Dec. 6-18		
Koonce, Albert F	Pvt	MG		W— Oct. 10-18	
Koonce, Walter G	Cpl	B		W— Oct. 17-18	
Koontz, George W	Pvt	G		W— Sept. 29-18	Nov. 6-18.
Koontz, Grover C	Sgt	K			
Koopman, Martin E	Pvt	H		W— Oct. 10-18	
Korte, Bernard	Pvt	K		K— Aug. 26-18	
Kotula, George	Pvt	H	Dec. 6-18		
Kowalczyk, Sam	Pvt	A			

Kowalkowski, Walenty	Pvt	H	Dec. 6-18		
Kowalzcyk, Waladyslaw	Pvt	H	Dec. 6-18		
Kramer, Benjamin H	Pvt	M			
Kraeka, Pietre	Pvt	H	Dec. 6-18		
Krawazyk, Tony	Pvt--1st Cl	C			
Kretzmier, Herbert C	Pvt	H	Dec. 6-18		
Krieg, Henry H	Pvt	H	Dec. 6-18		
Kroeger, Carl H	Pvt	H		K— Sept. 29-18	
Krohmalenke, Alexander	Pvt	H	Dec. 6-18		
Krononberger, Joseph S	Pvt	K			
Kronern, Reiholdt H	Pvt	A		W— Sept. 29-18	Dec. 7-18.
Kronick, Harry	Pvt	K		S— Mar 16-18	
Krucillock, John	Pvt	H	Dec. 6-18		
Kuberski, Zifert	Pvt	H	Dec. 6-18		
Kuebler, Otto E	Pvt--1st Cl	MD	Oct. 28-18		
Kuehn, Erwin F	Pvt	K		W— Aug. 31-18 K— Oct. 10-18	
Kuhlenengken, Honaman	Pvt	A		W— Oct. 17-18	Nov. 14-18.
Kuhlmann, William C	Pvt	H		K— Sept. 29-18	
Kyle, David E	Pvt	B			

Kyle, William B	Pvt	B			
Kyle, Granville E	Pvt	B			
Kyle, Samuel E	Pvt--1st Cl	Hq			
Labbee, Claude A	Cpl	G		W— Sept. 2-18 W— Sept. 29-18	Nov. 6-18.
Lacy, George J	Pvt--1st Cl	M			
Lachelt, Alfred H	Pvt--1st Cl	M		W— Oct. 10-18	
Ladd, William C	Pvt	E			
Lail, Jordan B	Pvt	M			
Lail, Luther L	Cpl	M			
Lail, James	Pvt	A		S— Nov. 27-18	
Laird, Leo C	Pvt	H	Dec. 6-18		
Lane, Charles F	Pvt--1st Cl	A		K— Sept. 3-18	
Lamagri, Louis	Pvt	H	Dec. 6-18		
Laney, John C	Pvt	A		K— Sept. 29-18	
Langford, Roby	Pvt	D		W— Oct. 27-18	
Langley, Henry	Pvt--1st Cl	Sup		S— Nov. 11-18	
Lamb, Fred E	Pvt	I			
Lamb, Frank L	Ck	MG			
Lamm, George H	Cpl	D			

Lamm, Herbert W	Pvt	Sup			
Lamm, Herbert L	Pvt	Sup			
Lamm, Pink	Pvt--1st Cl	K			
Lamm, Walter L	Sgt	A			
Lancaster, James A	Pvt	D			
Land, Fred	Pvt--1st Cl	F			
Land, George	Pvt	E			
Lane, Claude A	Cpl	A			
Lane, Hilmer T	Pvt	M			
Lane, Rester Z	Pvt	M			
Lane, James S	Cpl	A			
Lane, John T	Cpl	L			
Lang, Jack P	Sgt	I			
Langdon, Cornelius	Pvt	C		W— Sept. 29-18	Nov. 2-18.
Lange, Frederick G	Pvt	H	Dec. 6 -18		
Langley, James R	Ck	B			
Langston, Willie H	Cpl	F			
Lape, Orville D	Pvt	D			
Larson, Harry W	Pvt	K		W— Sept. 29-18	

Larson, Milton J	Pvt	F		S— Sept. 25-18	
Larson, William	Pvt	H		K— Sept. 29-18	
Lasley, James M	Pvt	MD		W— Sept. 29-18	
Lashley, Connie	Mech	M			
Lassiter, Charles	Mech	I			
Lassiter, James E	Pvt	M			
Lassiter, William J	Pvt	C			
Latimer, Clifton W	Pvt--1st Cl	MD			
Lauderdale, Don L	Pvt	H			
Laturell, Edward J	Pvt	K		W— Sept. 29-18	
Laue, Edward H	Sgt	C		W— Oct. 17-18	Nov. 26-18.
Lauer, Albert A	Pvt--1st Cl	D			
Laughter, Jabe T	Pvt	M			
Laurant, John J	Pvt	I		W— Sept. 29-18	Nov. 11-18.
Lavello, Johnnie	Pvt	MG	Oct. 26-18		
Lawing, Granham L	Pvt	F		W— Sept. 29-18	Jan. 7-19.
Lawrence, Glenn H	Pvt	I		W— Aug. 2-18	Aug. 31-18.
Lawson, Andrew J	Pvt--1st Cl	B			
Lawter, King	Pvt	H		W— Sept. 29-18	Nov. 16-18.

Lawrence, Richard E	Pvt	E		W— Sept. 29-18	
Lawrence Roy	Pvt	B		W— Sept. 29-18	
Lawson, Benjamin R	Pvt	B		T— Jul 13-18	
Leazer, Albert R	Pvt	D		W— Oct. 17-18	
Leazer, John E	Pvt	D		S— Oct. 1-18	
Leckey, Ralph L	Pvt--1st Cl	L			
Ledford, Beecher	Pvt--1st Cl	M			
Ledford, Isaac	Pvt	I	Dec. 6-18		
Ledford, Lawrence	Pvt	C			
Ledbetter, Shirley	Bug	B		K— Sept. 29-18	
Ledford, Josh	Pvt	G		T— May 4-18	
Lee, Robert E	Pvt	I		W— Oct. 10-18	
Lee, Walter	Mech	E		W— Sept. 29-18	
Lee, Wilton H	Pvt	M		W— Oct. 10-18	
Lee, Archie F	Wag	Sup			
Lee, Ben A	Pvt	G		W— Sept. 29-18	Oct. 17-18.
Lee, Herman	Pvt	B		W— Oct. 17-18	Nov. 27-18.
Lee, Milton R	Pvt--1st Cl	L			
Lee, Paul H	RSS	Sup			

Lee, Perry	Pvt	B			
Lee, Robert F	Pvt--1st Cl	L			
Leech, George D	Pvt--1st Cl	Hq			
Leet, Edgar B	Pvt	G			
Lefler, Martin L	Pvt--1st Cl	C		W— Oct. 17-18 T— Jan. 20-19	
Lemert, Milo	Sgt	G		K— Sept. 29-18	
Lemon, Wade H	Pvt	H		W— Oct. 10-18	
Le Master, Jack	Bug	L			
Lemert, Nathan	Sgt	G			
Lemon, Eugene E	Pvt--1st Cl	L		W— Sept. 29-18	Nov. 11-18.
Lemons, Willie	Pvt	F			
Leonard, Alfred B	Pvt	F		W— Sept. 29-18	Nov. 25-18.
Leonard, Raymond J	Pvt--1st Cl	H			
Leonard, Willie E	Pvt	H			
Lentz, John F	Cpl	L		W— Oct. 17-18	
Lewis, Amos C	Pvt	M		K— Oct. 19-18	
Lewis, Dewey	Cpl	F		W— Sept. 29-18	
Lewis, Lonnie M	Pvt--1st Cl	L		K— Sept. 1-18	
Lewis, William L	Pvt	L		W— Oct. 10-18	

Lewis, Charlie	Pvt	L		W— Sept. 29-18	Nov. 11-18.
Lewis, Dallas	Pvt--1st Cl	L		W— Sept. 29-18	Dec. 5-18.
Lewis, Dutch W	Pvt--1st Cl	F		W— Sept. 29-18	Nov. 6-18.
Lewis, Horace L	Pvt--1st Cl	L			
Lewis, John L	Pvt	K			
Lewis, Kincheon	Pvt	B			
Lewis, Murray E	Sgt	M			
Lewis, Robert F	Sgt	G		W— Oct. 10-18	Dec. 7-18.
Leshnock, Michael	Pvt	K	Dec. 6-18		
Leslie, Lock A	Cpl	F			
Leta, Salvatore	Pvt	I	Dec. 6-18		
Leutz, George J	Pvt	B			
Lichea, Alphonso	Pvt	I	Dec. 6-18		
Liebhober, Israel	Pvt	I	Dec. 6-18		
Lieving, Otto A	Pvt	I			
Lifrage, Samuel L	Sgt	B			
Ligette, John W	Sgt	B			
Lillibridge, Roy P	Pvt	MD			
Lindsey, Otis	Pvt	C			

Lilly, Elmer B	Pvt	H		D of Disease – Nov. 11-18	
Lineberger, Albert L	Cpl	H		K— Sept. 29-18	
Lineberger, Erroll W	Cpl	E		K— Sept. 29-18	
Lineberry, Cyrus W	Pvt--1st Cl	A		K— Sept. 29-18	
Link, Carl	Pvt	D		K— Jul 24-18	
Link, Ollie R	Ck	M			
Linkous, James F	Cpl	G		W— Sept. 29-18 K— Oct. 18-18	Oct. 15-18.
Linton, Samuel F	Pvt	MG			
Lippard, Glenn	Pvt	C		W— Aug. 31-18 W— Oct. 17-18	Nov. 14-18.
Little, Raney B	Pvt	M		K— Sept. 29-18	
List, Oma V	Pvt	I			
Little, Charlie T	Pvt	I		W— Sept. 29-18	Jan. 7-18.
Little, Henry W	Pvt	K		W— Oct. 10-18	Oct. 31-18.
Little, Sam	Wag	Sup			
Little, Walter A	Pvt	G			
Little, Watt	Pvt	F			
Littleton, Calvin	Pvt	B			

Livermann, Herbert	Sgt	A			
Lloyd, Luther S	Pvt	I		T— Feb. 22-19	
Loane, Charles D	Pvt	Hq		T— Feb. 24-19	
Lockamon, James T	Sgt	M			
Lockamy, Patrick J	Pvt	Hq			
Lockard, John E	Pvt	F			
Laftaas, Roy L	Pvt	K			
Logan, Isaac	Cpl	F		W— Oct. 10-18	
Logan, William C	Pvt	E		K— Jul 26-18	
Logsdon, Bernard J	Pvt	F		S— Oct. 29-18	
Logsdon, Walter A	Pvt	M			
Logue, Charles H	Pvt	M			
Long, Leonard J	Pvt--1st Cl	M		K— Oct. 10-18	
Long, Charles G	Pvt	MD			
Long, Edward L	Pvt	K		W— Oct. 10-18	Dec. 5-18.
Long, Emory R	Pvt--1st Cl	C			
Long, Empie C	Cpl	MG			
Long, Kenneth L	Pvt--1st Cl	K			
Long, Paul A	Mus--2nd Cl	Hq			

Lonzo, Baldossoro	Pvt	A	Dec. 6-18		
Loomis, Orland	Pvt	K			
Loos, John A., Jr	Pvt	L		T— June 9-18	
Love, Alfred	Pvt	L			
Lovell, George R	Pvt	D			
Loven, George C	Cpl	F		W— Sept. 29-18	
Lowe, Roger Q	Sgt	B		K— Sept. 29-18	
Lowe, William E	Pvt--1st Cl	L		W— Sept. 29-18	
Lowery, John D	Pvt	M		W— Sept. 29-18	Sept. 29-18.
Lowory, John F	Pvt--1st Cl	MG			
Loyd, Thomas C	Pvt--1st Cl	A		W— Oct. 17-18	Feb. 25-19.
Lucas, Clarence L	Pvt	Sup			
Lucas, William M	Pvt	M			
Lufsey, Charles	Pvt--1st Cl	L		T— Feb. 8-19	
Lumpkin, Charles M	Cpl	G		W— Sept. 29-18	Dec. 19-18.
Lundberg, Gustav A	Pvt	F		W— Sept. 29-18	Dec. 8-18.
Lunte, Christian W	Cpl	A			
Lupton, Clyde E	Cpl	H		W— Aug. 11-18	Dec. 2-18.
Lyman, Asa	Pvt	K			

Lynch, Harley	Pvt	E		W— Sept. 29-18	Jan. 1-19.
Lutter, Zach T	Pvt	E		W— Sept. 29-18	
Lyders, Oscar	Mus	Hq		T— Feb. 16-19	
Lynch, Edward A	Pvt	H		K— Sept. 29-18	
Lyons, Leslie C	Pvt	E		W— Oct. 18-18	
Lyons, Ralph A	Bug	MG			
Lyons, George F	Pvt--1st Cl	Hq			
Lyons, Ira	Cpl	L			
Lyons, Thomas	Ck	L			
Mabry, Crawford L	Pvt--1st Cl	Sup		S— Nov. 11-18	
Mabe, James G	Pvt	A		W— Sept. 29-18	Nov. 6-18.
Mack, Thomas A	Pvt	Inf			
Macon, Thomas F	Pvt	G		W— Sept. 29-18	
Macy, Willie	Pvt	H		K— Sept. 29-18	
Madden, John T	Pvt	K		W— Oct. 10-18	
Maddox, George W	Pvt	L			
Madenjian, Kersam	Pvt	K	Dec. 6-18		
Madison, Clyde C	Pvt	Hq		S— Sept. 28-18	
Maffucci, Michael	Pvt	K	Dec. 6-18		

Mahan, William H	Pvt	L		W— Sept. 10-18	
Majcher, Anton	Pvt	K	Dec. 6-18		
Majette, Paul W	Pvt	Inf			
Majkowski, Boleslaw	Pvt	L		W— Sept. 29-18	Oct. 17-18.
Majors, Ira	Pvt--1st Cl	Hq			
Malkin, Max	Pvt	Inf		T— Jul 17-18	
Mallozzi, Antonio	Pvt	K	Dec. 6-18		
Mangum, Callahan	Pvt	M			
Mangum, Harrison F	Pvt	D		K— Sept. 29-18	
Manley, William H	Wag	Sup		W— Oct. 7-18	
Maples, Oscar	Pvt	L		W— Oct. 10-18	
Maranville, Clyde D	Pvt	L		S— Sept. 30-18	Dec. 2-18.
Marchiana, Frank	Pvt	I			
Marcozzi, Sam	Pvt	K	Dec. 6-18		
Maready, Gaylor	Pvt--1st Cl	F			
Margulis, Michael I	Pvt	L		K— Aug. 31-18	
Mariele, James	Ck	Hq		S— Nov. 11-18	
Marineau, Ernest	Pvt	K	Dec. 6-18		
Marker, Guy E	Pvt	M		S— Oct. 31-18	Dec. 14-18.

Marlow, Claudis B	Pvt	K			
Marlow, Charles R	Pvt	K			
Marlow, Eli M	Pvt--1st Cl	Sup			
Marlow, Lee	Pvt--1st Cl	Sup			
Marquard, Henry M	Pvt	G			
Marrone, Marine	Pvt	K	Dec. 6-18		
Marshall, Andy	Cpl	H		S— Oct. 10-18 T— Jan. 3-19	Oct. 28-18.
Marshall, George N	Pvt	E			
Marshall, Joseph W	Pvt	Hq			
Marshall, Watter	Pvt--1st Cl	F		W— Sept. 29-18	Dec. 2-18.
Marshall, Coit O	Sgt	H		W— Sept. 29-18 Jan. 20-19	Oct. 29-18.
Marshall, Aquilla B., Jr	Cpl	H		T— Jan. 22-19	
Marshburn, Jacob A	Pvt	MG		K— Sept. 29-18	
Martin, Frank P	Pvt	L		S— Oct. 30-18	
Martin, James A	Pvt	E		T— Jul 12-18	
Martin, John	Pvt--1st Cl	L		W— Oct. 17-18	
Martin, Codie	Pvt	Inf			
Martin, Daniel W	Pvt	K			

Martin, Edward D	Pvt	F		M— Sept. 29-18	Dec. 7-18.
Martin, Fred	Pvt--1st Cl	E		W— Oct. 17-18	Dec. 3-18.
Martin, George E	Cpl	E			
Martin, Henry M	Pvt--1st Cl	G		S— Sept. 29-18	Nov. 26-18.
Martin, Hugh B	Cpl	MG		S— Oct. 25-18	Dec. 2-18.
Martin, James E	Pvt	F			
Martin, Jeanery	Pvt--1st Cl	O		S— Nov. 11-18	Dec. 1-18.
Martin, Leslie	Pvt--1st Cl	L			
Martin, Rodus	Sgt	L			
Martin, William R	Pvt--1st Cl	Hq			
Martin, William M	Pvt	M		W— Oct. 10-18	Dec. 14-18.
Martinez, Simon	Pvt	K	Dec. 6-18		
Martole, Dominick	Pvt	L			
Marzello, Saverio	Pvt	K	Dec. 6-18		
Masaitis, George	Pvt	K	Dec. 6-18		
Mashburn, Charles E	Pvt	M			
Masi, Pietro	Pvt--1st Cl	Inf			
Mason, John T	Pvt	F			
Mason, Dosie E	Pvt--1st Cl	G		W— Sept. 29-18	

Mason, Samuel	Pvt	L		K— Sept. 29-18	
Massey, James R	Pvt	L			
Massey, John R	Pvt	C		W— Sept. 3-18	Dec. 7-18.
Massey, William O	Cpl	C		W— Oct. 17-18	Dec. 2-18.
Massie, Eddie C	Pvt--1st Cl	F		W— Sept. 29-18	Dec. 3-18.
Massingill, Fred	Bug	M			
Masten, Charles	Mus	Hq			
Mastin, Thomas E	Pvt--1st Cl	M			
Mathews, Ernest L	Cpl	M			
Mathews, Troy L	Pvt	MG			
Mathias, William N	Pvt	H		S— Aug. 31-18	Nov. 1-18.
Matkins, William H	Pvt	B			
Matthews, Charlie A	Pvt--1st Cl	D		W— Sept. 29-18	Nov. 11-18.
Matthews, James E	Pvt--1st Cl	C		W— Sept. 29-18	Dec. 7-18.
Matthews, John C	Pvt	Hq			
Matthews, Joseph M	Pvt	M		W— Oct. 18-18	Nov. 11-18.
Matthews, George W	Pvt	M		Oct. 10-18	
Matthews, Kenneth C	Mess Sgt	M		T— Jan. 16-19	
Matthews, Milton A	Pvt	M		K— Oct. 19-18	

Matthews, Thomas M	Pvt	F		W— Sept. 29-18	
Mattox, Bryant	Pvt	M		W— Aug. 26-18	
Mauney, George M	Pvt--1st Cl	Hq		W— Sept. 29-18	Nov. 26-18.
Maxwell, Clifford	Pvt	F		S— Nov. 1-18	Dec. 7-18.
Maxwell, William C	Pvt	E			
May, Willie B	Mech	E			
May, Rudy E	Pvt	Hq		S— Nov. 1-18	
Mayes, Clarence	Pvt	D			
Maynard, Jacob	Pvt	H			
Maynard, Silas T	Pvt	A		S— Aug. 31-18	Sept. 15-18.
Maynor, Wilburt	Pvt	H			
Mayta, Martin T	Pvt	F			
Mazzini, Alphonso	Pvt	A	Dec. 6-18		
McAdams, Archie A	Pvt	G			
McAnnaly, John A	Pvt	E		T— May 1-18	
McBaer, Clyde	Pvt	Inf			
McBay, Walter	Cpl	E		W— Sept. 29-18	Oct. 29-18.
McBay, Will	Pvt	E		W— Sept. 29-18	
McBee, Charlie	Cpl	L			

McBrayer, Charles S	H.S.H.R	Sup			
McCall, Lawton	Cpl	MG			
McCall, Roy	Pvt	E		S— May 1-18	
McCann, Elzia	Pvt	H		W— Sept. 29-18	Dec. 2-18.
McCarkle, Ralph	Wag	Hq			
McCaskill, Dann	Pvt	C			
McCaslin, Mort	Pvt	M		W— Sept. 29-18	
McCausley, Zennie	Pvt	M			
McClaren, Oliver W	Cpl	M			
McClenney, Robert L	Pvt	I			
McCluskey, John J	Pvt--1st Cl	M			
McCollum, Watson S	Sgt	L		W— Oct. 17-18	
McCool, Edward	Pvt--1st Cl	E			
McCormac, James E	Cpl	L		T— Aug. 1-18	
McCormack, Duncan	Pvt	H			
McCormick, Michael L	Pvt	Ind			
McCoy, Herbert E	Sgt	L			
McCray, Coleman A	Sgt	G		K— Jul 26-18	
McCrory, Roy D	Cpl	C		S— Oct. 1-18	Oct. 27-18.

McCullen, Erie	Sgt	MG		T— Jan. 29-19	
McCullough, Charles B	St. Sgt	Hq			
McCullough, John F	Cpl	Hq		W— Sept. 29-18	Dec. 1-18.
McCullough, Ernest	Pvt	H		K— Sept. 29-18	
McCullough, William W	Cpl	E		W— Sept. 29-18	
McCurdy, Carrie W	Pvt	MG			
McDevett, Dennis P	Pvt--1st Cl	Hq			
McDevett, Frazier T	Sgt	Hq		W— Sept. 29-18	
McDaniel, Paul M	Pvt--1st Cl	Hq			
McDine, Richard	Pvt	F			
McDonald, Allen	Pvt	M		W— Aug. 5-18	Nov. 26-18.
McDonald, Richard L	Pvt--1st Cl	M			
McDonald, Benjamin A	Sgt	L			
McDonald, Robert	Pvt	M		W— Oct. 10-18	Dec. 14-18.
McDonald, Arthur L	Pvt	L		W— Aug. 31-18	
McDonald, Eugene	Cpl	F		W— Sept. 29-18 T— Jan. 20-19	Dec. 7-18.
McDonald, John R	Pvt 1st Cl	G		T— Jul 18-18	
McDonald, John T	Cpl	D		W— Oct. 17-18	
McDougald, Charles A	Sgt	A		K— Jul 22-18	

McDowell, William J	Pvt	Inf		T— Jan. 29-19	
McDuffie, William H	Wag	Sup			
McDuffie, Paul A	Sgt	MG		S— Oct. 14-18	Jan. 19-19.
McElreath, Leonard B	Pvt	Sup			
McGee, Edgar L	Cpl	M			
McGee, Joseph A	Wag	Sup		W— Oct. 7-18	
McGee, Julius M	Pvt	F			
McGehee, Thomas A	Pvt	E		W— Oct. 17-18 T— Jan. 22-19	Nov. 6-18.
McGill, Peter T	Pvt	Hq			
McGilvary, John A	1st Sgt	F			
McGougan, Ernest D	Sgt	L			
McGowan, Hester C	Mech	MG		D— Dec. 23-18	
McGuinn, Willie E	Pvt	M			
McGuire, Roy	Pvt	K			
McGuirt, Charles B	Pvt--1st Cl	A		W Oct. 17-18	Dec. 7-18.
McIntosh, Otis L	Cpl	L			
McIntosh, Carl	Pvt--1st Cl	M		S— Oct. 31-18	
McIntyre, Haywood S	Pvt--1st Cl	F			

McIntyre, Daniel A	Pvt	M		S— Oct. 31-18	Dec. 3-18.
McKay, William	Pvt	E		S— Jan. 7-19	Jan. 19-19.
McKaoun, Verner	Pvt	M			
McKee, Willie	Pvt	F		S— May 1-18	
McKeel, James S	Sgt	K			
McKelvey, Sheeler Z	Pvt	F			
McKenzie, George R	Pvt--1st Cl	MG		S— Dec. 1-18	
McKinney, John	Pvt--1st Cl	M			
McKinney, Stokes	Pvt	M		W— Oct. 10-18	Oct. 27-18.
McKinney, Homer	Pvt--1st Cl	M		D— Sept. 23-18	
McLain, Alex M	Pvt	K			
McLain, Kenneth S	Cpl	A			
McLain, Allen B	Pvt	B		K— Oct. 18-18	
McLain, Harry C	Pvt	L		Deserted Sept. 27-18	
McLaren, Ray	Pvt	F		W— Sept. 29-18	Nov. 8-18.
McLaughlin, Henry N	Pvt--1st Cl	MG		W— Jul 25-18	
McLaughlin, John H	Cpl	F		W— Sept. 28-18	Dec. 3-18.
McLaughlin, Milton	Pvt	M		K— Sept. 29-18	
McLaughlin, Erenest D	Cpl	Sup			

McLaurin, Harvey E	Pvt--1st Cl	A		W— . 29-18	
McLaurin, William	Mech	H		W & D— Oct. 14-18	
McLean, Homer C	Pvt	H		W— Sept. 29-18	Dec. 2-18.
McLean, John N	Pvt--1st Cl	M			
McLester, Charles	Pvt--1st Cl	MG			
McMillan, Charles G	Pvt	I		K— Sept. 29-18	
McMillan, William E	Pvt	H			
McNair, Harvey C	Cpl	K		W— Sept. 20-18	Dec. 14-18.
McNair, Howard C	Pvt	L			
McNaughton, Fred W	Pvt	C		W— Oct. 10-18	Dec. 12-18.
McNeil, Frank E	Pvt	E		M— Oct. 17-18	
McNeil, Luther M	Pvt--1st Cl	L		W— Oct. 17-18 T— Feb. 28-19	Dec. 14-18.
McNeill, Kenneth	Sgt	F			
McNeill, Clarence	Sgt	L			
McNutt, Glen A	Pvt	L		S— Nov. 1-18	
McPail, John D	Cpl	L		K— Sept. 29-18	
McPhail, Colon	Pvt--1st Cl	M		W— Aug. 5-18	Aug. 5-18.
McPherson, Charles H	Cpl	M		K— Oct. 10-18	
McQueese, Robert	Pvt	Sup			

McSwain, Baxter C	Pvt--1st Cl	M		W— Sept. 29-18	
McSwain, Elezer	Wag	Sup			
Meadows, Guy S	Pvt	Hq			
Meadows, Robert R	Sgt	Hq			
Meakin, Douglas H	Pvt	H		W— Sept. 29-18	
Meaks, Braxton O	Pvt	M		W— Oct. 17-18	
Means, Edward L	Pvt--1st Cl	M			
Mears, William R	B. Sgt	Hq			
Meeks, William J	Pvt	I			
Mehaffey, Jerry C	Sgt	F		S— May 1-18	
Melson, William J	Cpl	H		W— Sept. 29-18	Nov. 11-18.
Melting, Oliver	Cpl	H			
Melton, William X	Pvt	L		K— Aug. 2-18	
Melvin, Colon	Pvt	I		W— Aug. 31-18	
Melvin, Marley M	Sgt	F		W— Oct. 10-18 T— Jan. 20-19	Nov. 14-18.
Melvin, Claude G	Pvt--1st Cl	MG			
Melvin, Edward G	Pvt--1st Cl	G			
Melte, Frank O	Pvt	M		M— Oct. 10-18	Jan. 22-19.
Menzeke, Charles E	Pvt	K		K— Aug. 26-18	

Meredith, Marshall S	Cpl	E		W— Oct. 10-18	
Meredith, Charles L	Pvt	K			
Merritt, George T	Pvt	M			
Merritt, Jesse C	Pvt	F			
Merritt, John B	Pvt	F			
Merrow, Flavious J	Pvt--1st Cl	E		K— Sept. 29-18	
Messer, Isaac H	Pvt--1st Cl	L			
Messer, William E	Pvt	C		W— Oct. 17-18	Nov. 26-18.
Mette, John A., Jr	Pvt--1st Cl	E		W— Sept. 29-18	Dec. 7-18.
Metts, George L	Pvt	B		W— Sept. 29-18	
Metzger, Orville C	Pvt	K		T— June 1-19	
Mianogich, Stanke	Cpl	L			
Michael, Odie	Pvt--1st Cl	G		T— Jan. 22-19	
Mickey, John A	Pvt--1st Cl	A		K— Oct. 10-18	
Mickey, James	Pvt	E			
Mickey, William L	Pvt	H		S— Oct. 29-18	Dec. 21-18.
Mickle, Harold E	Pvt	M		S— Oct. 31-18	
Mickle, Herbert L	Pvt	K		W— Sept. 29-18	Dec. 2-18.
Midyette, Payne H	1st Sgt	B		T— Sept. 10-18	

Milay, Henry J	Pvt	M		W— Aug. 31-18	
Milenskas, Joseph	Pvt	K	Dec. 6-18		
Miles, Robert R	Pvt	Hq		S— Nov. 11-18	Nov. 26-18.
Miles, Walter	Pvt	L			
Milholland, Robert S	Cpl	Hq			
Millard, William F	Pvt	H		W— Sept. 29-18	Dec. 2-18.
Miller, Paul L	Pvt	Hq			
Miller, Bennett W	Pvt	A			
Miller, Clinton F	Cpl	L			
Miller, Gus V	Pvt	H			
Miller, Hobart	Pvt--1st Cl	G			
Miller, Roy	Pvt	H			
Miller, Thomas J	Pvt--1st Cl	M			
Miller, William W	Pvt	I			
Miller, Zed T	Pvt	I		S— Nov. 13-18	Nov. 27-18.
Miller, Craig	Cpl	G		T— Jul 15-18	
Miller, Gaither	Pvt	G		K— Sept. 27-18	
Miller, George F	Sgt	K		W & D— Sept. 29-18.	
Miller, George W	Cpl	K		W— Oct. 10-18	

Miller, Glenn W	Sgt	H		W— Sept. 29-18	
Miller, John T	Pvt	G		K— Oct. 18-18	
Miller, Samuel	Sgt	H		W— Sept. 29-18	
Mills, Elgie	Pvt	E		K— Sept. 29-18	
Mills, Raymond	Pvt	H		W— Oct. 10-18	
Michael, William H	Pvt	H		W— Sept. 29-18	Oct. 25-18.
Mick, Horace	Pvt	B			
Millis, Edward P	Pvt	M			
Mills, Fred R	Pvt	Hq			
Mills, James L	Pvt--1st Cl	G			
Mills, Mosie I	Pvt--1st Cl	D			
Mills, Stamey	Pvt	Hq			
Mills, William A	Pvt	F			
Milton, Charles C	Pvt	Inf			
Minns, William A	Pvt	I		W— Sept. 29-18	
Minter, Edwin J	Pvt	F			
Mintz, William A	Pvt--1st Cl	A			
Minter, Lennie L	Pvt--1st Cl	F			
Minton, John	Pvt	F		W— Sept. 29-18	Dec. 7-18.

Mintz, Van G	1st. Sgt	E		W— Oct. 10-18	Nov. 1-18.
Mitchell, William R	Pvt--1st Cl	D		W& D— Oct. 11-18.	
Mitchell, James M., Jr	Sgt--1st Cl	Inf			
Mitchell, Jasper H	Pvt--1st Cl	C			
Mitchell, Robert R	Cpl	G		W— Oct. 10-18	Dec. 2-18.
Mitchell, William M	Pvt	G			
Mizinski, Wladyslav	Pvt	A		K— Oct. 17-18	
Mizzer, Stanley G	Pvt	G		W— Sept. 29-18	
Mobley, Charlie R	Sgt	F			
Modderman, Charlie	Pvt	C			
Modlin, William	Pvt	D		S— May 1-18	
Moffet, Ervin B	Cpl	M			
Mohr, Earl C	Pvt	E		S— Nov. 22-18	Dec. 1-18.
Monday, James R	Pvt--1st Cl	G		W— Sept. 23-18 T— Jan. 22-19	Oct. 27-18.
Monroe, Edward C	Pvt	F			
Monroe, Ernest H	Pvt--1st Cl	M		S— Oct. 31-18	Nov. 26-18.
Monson, George E	Pvt 1st Cl	D			
Monsteller, Wilburn	Pvt--1st Cl	K			
Moon, John E	Pvt	K			

Mooney, George	Pvt	A			
Moore, Asbury H	Pvt--1st Cl	L		W— Oct. 17-18	Dec. 14-18.
Moore, Carl	Cpl	D		W— Sept. 29-18	Nov. 11-18.
Moore, Clarence M	Wag	Sup			
Moore, Elliott L	Bugler	K			
Moore, Frank A	Pvt	K		S— Oct. 31-18	Nov. 26-18.
Moore, George H	Cpl	A		W— Oct. 17-18	Nov. 26-18.
Moore, George W	Pvt	Hq		W— Aug. 25-18	Dec. 1-18.
Moore, Harold H	Pvt--1st Cl	D			
Moore, Herman	Pvt	B		W— Sept. 29-18	Dec. 14-18.
Moore, John W	Pvt	M			
Moore, Joseph H	Pvt--1st Cl	L		M— Aug. 31-18	Dec. 29-18.
Moore, Lloyd B	Pvt	F			
Moore, Samuel A	Sgt	L		S— Oct. 30-18	Nov. 4-18.
Moore, Shelton W	Sgt	I			
Moore, Thomas	Cpl	E		W— Sept. 29-18	Nov. 11-18.
Moore, Wade	Pvt	I			
Moore, Walter H	Pvt--1st Cl	F			
Moore, Walter T	Pvt	L		W— Oct. 17-18	Dec. 21-18.

Moore, Warner E	Mech	C			
Moore, Albert	Pvt	F		W— Sept. 29-18 W— Oct. 17-18	
Moore, Charles T	Pvt--1st Cl	M		W— Sept. 29-18	
Moore, Emmitt H	Cpl	K		K— Sept. 29-18	
Moore, Grover C	Pvt	O		S— Mar 15-18	
Moore, Lonnie M	Pvt	D		W— Sept. 29-18	
Moore, Phillip L	Pvt	M		W— Oct. 10-18	
Moore, Robert A	Cpl	K		W— Sept. 29-18	
Moorehead, Alvin	Sgt	H		T— Feb. 4-18	
Morecock, Herbert	Pvt	E			
Morgan, Charles G	Pvt--1st Cl	D		S— Oct. 29-18	Dec. 10-18.
Morgan, Clifford J	Pvt--1st Cl	Hq		S— Aug. 31-18	Oct. 20-18.
Morgan, Emmet L	Cpl	I		W— Oct. 10-18	Dec. 2-18.
Morgan, John C	Pvt	K		W— Sept. 29-18	Dec. 2-18.
Morgan, Logan	Pvt	D			
Morgan, Robert W	Pvt	Inf		W— Sept. 29-18	Oct. 28-18.
Morgan, Sam J	Pvt--1st Cl	B			
Morgan, Whitney W	Pvt	D	Dec. 6-18		
Morgan, Whitney J	Sgt	Hq			

Morgan, Lamar	Pvt	B		K— Jul 18-18	
Morgan, Robert W	Cpl	G		S— Mar. 15-19	
Morgan, Roy J	Pvt	M		K— Oct. 19-18	
Morrie, Harry R	Pvt	L		T— Feb. 8-19	
Morris, Overton T	Cpl	A		K— Sept. 29-18	
Morris, Walter	Pvt	M		W— Oct. 10-18	
Morris, Charlie H	Pvt--1st Cl	K		W— Sept. 29-18	Nov. 26-18.
Morris, Covie	Pvt	A			
Morrison, Leroy	Pvt	F		S— Nov. 1-18	
Morrow, Jeff L	Pvt	F			
Mort, James K	Pvt	F			
Mortin, Leo W	Sgt	K		T— Jan. 27-19	
Morton, Will	Pvt	E		S— May 1-18	
Moser, Benton B	Cpl	E		S— Nov. 11-18	Dec. 3-18.
Moser, Eugene E	Pvt--1st Cl	C		M— Sept. 29-18	
Moss, James B	Sgt	K		W— Sept. 29--18	
Moss, Ernest	Pvt	E			
Moss Henry	Pvt	F			
Moss, George W	Sgt	I			

Moss, Jesse L	Pvt	B			
Moss, Wilford H	Pvt--1st Cl	Hq			
Motsinger, Chester	Pvt	K		S— Nov. 29-18	
Moye, Thaddeus	Sgt	MG			
Moyers, James R	Cpl	E			
Moyers, James W	Sgt	E			
Moyers, Rexie E	Cpl	E		W— Sept. 29-18	
Mozingo, Norman	Pvt--1st Cl	C		S— Aug. 3-18	Sept. 12-18.
Muelken, Martin H	Pvt	A			
Mullis, Boston	Pvt--1st Cl	F			
Mullen, Thomas L	Pvt	Inf			
Mulherin, John R	Pvt--1st Cl	K		K— Sept. 29-18	
Mullin, Willie G	Cpl	A		W— Sept. 29-18	
Mulinax, Charlie F	Pvt	I		W— Sept. 29-18	
Mullinix, Franklin M	Pvt--1st Cl	B		W— Oct. 17-18	
Mullins, Sam	Pvt	H		W & D— Sept. 29-18	
Mumford, Charlie J	Sgt	K			
Munday, Morgan O	Pvt	H		W— Sept. 29-18	Dec. 7-18.
Munday, Ellis	Pvt	H			

Munday, Benjamin H	Pvt	A		W— Oct. 17-18	
Munden, Bernard C	Pvt--1st Cl	K		W— Sept. 29-18	Dec. 2-18.
Munie, Raymond C	Pvt	K		S— Dec. 29-18	
Munns, Eunice	Pvt	D		W— Sept. 29-18	Dec. 8-18.
Murphy, Harold F	Pvt	G			
Murphy, James F	Pvt--1st Cl	D		W— Sept. 29-18	Dec. 20-18.
Murphy, Thomas J	Pvt	M			
Myers, Charlie	Pvt	C			
Myers, Frank E	Pvt	D			
Myers, Raymond W	Pvt	M			
Myers, Rufus	Cpl	M			
Myers, Columbus	Pvt--1st Cl	M		W— Oct. 10-18	
Myers, Homer L	Pvt	M		S— Oct. 20-18	
Myers, James M	Pvt	I		S— Sept. 29-18	
Murphy, Will	Pvt	C		W— Sept. 29-18	
Musgrave, Bruno	Pvt	MG		W— Sept. 29-18	
Myles, Robert H	Pvt--1st Cl	Hq			
Myrland, Richard G	Pvt	Inf			
Nanee, Alvah H	Ck	I			

Nance, Charles D	Pvt	Hq			
Nance, Claude I	Pvt	Hq			
Narron, James W	Pvt--1st Cl	C		W— Oct. 17-18	Nov. 20-18.
Narveson, Palmer O	Sgt	H			
Nash, Charlie E	Pvt	L			
Nash, William C	Cpl	G			
Nation, Jesse	Pvt	F		W— Sept. 29-18	Oct. 17-18.
Neal, Evans	Pvt--1st Cl	Hq			
Neal, Foster A	Pvt	Sup		W— Oct. 27-18	
Neal, Thomas J	Pvt	E		K— Sept. 29-18	
Neblett, Robert W	Cpl	Hq			
Needleman, Samuel J	Pvt	I			
Neeley, James A	Cpl	E			
Neely, James L	Pvt	L		W— Oct. 17-18	
Nelson, David	Cpl	B		W— Sept. 29-18	Dec. 2-18.
Nelson, Dent K	Pvt	F		W— Sept. 29-18	
Nelson, Julius T	Pvt--1st Cl	MD		T— Nov. 2-18	
Nelson, Leo H	Pvt	G		W— Oct. 17-18 T— Jan. 22-19	Nov. 2-18.
Nelson, Ludwig	Pvt	Hq			

Nelson, Neal A	Cpl	Hq			
Nelson, Reuben F	Col. Sgt	Hq			
Nelson, Willard	Pvt	M			
Nesbit, Claude W	Ck	E			
Nesbitt, Albert	Pvt	K		S --Oct. 20-18	
Newell, Herbert W	Pvt	G		S— Oct. 29-18	
Newman, William R	Pvt	B		K— Oct. 18-18	
Newberry, William C	Pvt	G			
Newell, Fred T	Cpl	E		W— Sept. 29-18	Dec. 2-18.
Newlin, Jacob L	Pvt	A		W— Sept. 29-18	Nov. 26-18.
Newman, John S	Pvt	B			
Newsom, Ellis F	Pvt	D			
Newson, Neuman W	Mech	Hq			
Newsom, Brady H	Pvt	H		W— Sept. 29-18	
Newsom, Windsor T	Pvt	MG		T— Jan. 20-19	
Newton, John	Wag	Hq		T— Jan. 20-19	
Newton, Clifton	Pvt--1st Cl	I			
Newton, Issac M	Cpl	H			
Newton, Jesse M	Cpl	Hq			

Nichols, Tony	Pvt	L			
Nicholson, Frank C	Pvt	K			
Nicks, Jesse	Pvt	H		W— Oct. 10-18	Nov. 2-18.
Niell, Leonard R	Cpl	H		S— Oct. 29-18	
Nimocks, David R	Cpl	Hq		T— Jul 22-18	
Nitsche, William J	Mus	Hq	Nov. 14-18		
Nixon, Charles	Cpl	K			
Nixon, Rosser G	Pvt	M			
Noble, Archie	Cpl	I			
Noble, Frank H	Pvt	Hq			
Nogrotski, William	Pvt	K			
Nolan, Thomas A	Pvt	Hq			
Noles, Lonnie	Pvt	M			
Nollen, Matthew	Pvt--1st Cl	B		T— Sept. 12-18	
Nordmeyer, Henry	Pvt	C			
Norem, Martin E	Pvt	MD			
Norfleet, Joseph E	Pvt--1st Cl	L		T— Feb. 8-19 W— Oct. 10-18	Dec. 2-18.
Norman, Joseph	Pvt	E		Sept. 29-18	Dec. 2-18.

Noris, Mack	Pvt	C		T— May 1-18	
Norman, Isaac C	Pvt	A		K— Sept. 29-18	
Norman, William C	Pvt	F		T— May 1-18	
Norris, Edgar	Pvt--1st Cl	M			
Norris, Ernest	Pvt	M		C— Oct. 11-18	Dec. 8-18.
Norris, Jimmie G	Pvt--1st Cl	G		K— Sept. 29-18	
Norris, Louis H	Pvt--1st Cl	F			
Norris, Zell	Pvt--1st Cl	B			
Norris, Loy	Cpl	Hq			
Norrod, Ernest L	Cpl	B			
Norrod, Floyd J	Cpl	G		W— Sept. 29-18	Nov. 11-18.
Norton, Ben	Pvt--1st Cl	C		W- Oct. 10-18	
Norton, John R	Cpl	K			
Nowakowski, Joseph J	Pvt	L		W— Aug. 31-18	Nov. 18-18.
Nowtnh, Arthur T	Pvt	K		W— Sept. 29-18	Dec. 5-18.
Nowotny, John	Pvt	Hq			
Nunn, Rudolph	Cpl	Hq			
Oakley, Richard	Pvt--1st Cl	I			
Oakley, Water A	Pvt--1st Cl	B		W— Oct. 17-18	

Oates, John L	Pvt	MG			
Oates, Howard G	Cpl	D		K— Sept. 29-18	
O'Brien, Otis B	Cpl	D		K— Oct. 14-18	
Obarowski, Frank A	Pvt	K	Dec. 6-18		
Obar, Cliff	Cpl	K			
O'Brine, James B	Pvt--1st Cl	M		W— Oct. 10-18	Dec. 14-18.
O'Brine, William L	Pvt	M		W— Sept. 29-18	Nov. 22-18.
O'Danicl, Stephen	Sgt	Hq			
Oguin, Cad B	Pvt	L		S— Nov. 11-18	Jan. 3-19.
Oldham, Frank A	Pvt	E		T— Jan. 22-19	
Oliver, Willard	Pvt--1st Cl	D		T— Feb. 4-19	
Olrikson, Mads A	Sgt	D			
Olson, Bert	Pvt	F		W— Sept. 29-18	Dec. 19-18.
Olson, Carl A	Pvt	K	Dec. 6-18		
Olson, Jacob E	Pvt	A			
Olson, Odin	Pvt	MG			
Olson, Oscar B	Pvt	MG			
Oller, John	Pvt	H		W— Sept. 29-18	
Olson, Nels F	Pvt	A		K— Aug. 17-18	
Olson, Anton	Pvt	C		T— May 4-18	

O'Malley, Charles S	Pvt	L		W— Aug. 31-18	Nov. 14-18.
O'Meara, Charles A	Sgt	Sup			
O'Neal, Samuel R	Pvt	M			
O'Neal, Thomas J	Sgt	C		Discharged— Jul 1-18	
O'Neill, William B	Pvt	A		S— Oct. 28-18	
Osborne, Jesse A	Pvt	G		W— Sept. 29-18	
Ore, Brice	Wag	Sup			
Orloske, Albert B	Pvt	A			
Osga, John J	Pvt	G		W— Sept. 29-18	
Oswald, Lawrence J	Pvt	L		S— Oct. 3;-18 T— Feb. 8-19	
Oswalt, Rollin	Cpl	C		W— Oct. 17-18!	Dec. 2-18.
Ott, Robert	Pvt	M			
Ott, Daid	Pvt--1st Cl	L		K— Sept. 29-18	
Otterstrom, Kjalmar	Pvt	C			
Owen, Paul C	Cpl	C		T— Jan. 8-18	
Owen, Roland	Pvt	D		T— Dec. 17-18	
Owens, Bernice C	Cook	H		T— Jul 12-18	
Owens, Ellis A	Pvt	I		D of W— Aug. 2-18	
Overman, Thomas M	Pvt--1st Cl	MG			

Owens, Bernice L	Pvt--1st Cl	K			
Owens, Herman F	1st Sgt	C		W— Oct. 17-18	
Owens, Jesse L	Pvt	K		W— Sept. 29-18	Dec. 8-18.
Owensby, John D	Pvt	D		W— Oct. 10-18	
Pack, John I	Pvt	Hq			
Padgett, Critchley	Pvt	F		S— Oct. 14-18	Dec. 7-18.
Page, Israel	Pvt	F		W— Sept. 29-18	Oct. 27-18.
Palmer, William H	Pvt	Sup			
Palmer, Charles F	Pvt	B		W— Oct. 17-18	
Palmer, Dee D	Pvt	G		W & D --Sept. 29-18	
Pals, Hilko	Pvt	Inf			
Parham, Albert	Ck	F		W— Oct. 10-18	Dec. 3-18.
Park, Fred	Pvt	H			
Parker, John W	Pvt--1st Cl	A		K— Oct. 17-18	
Parker, Samuel E	Pvt	Sup		S— Nov. 11-18	
Parker, Alfred	Pvt--1st Cl	E		W— Oct. 17-18	Dec. 3-18.
Parker, Allcye	Bglr	B		S— Nov. 11-18	Nov. 27-18.
Parker, Charlie C	Wag	Sup			
Parker, Clarence	Pvt--1st Cl	I			
Parker, Eugene	Pvt--1st Cl	I			

Parker, Harrison	Pvt	M			
Parker, Herbert L	Mess Sgt	M			
Parker, Hunter A	Pvt--1st Cl	M			
Parker, Jasper Y	Pvt	Y	Dec. 6-18		
Parker, Jasse B	Pvt	H			
Parker, John C	Pvt--1st Cl	Inf			
Parker, Lonnie C	Wag	Sup		W— Oct. 7-18	Nov. 11-18.
Parks, Samuel W	Sgt	Sup			
Parrish, Donnie C	Pvt--1st Cl	C			
Parrish, John B	Cpl	Hq			
Parrish, Walter	Pvt--1st Cl	A		W— Sept. 3-18	Dec. 2-18.
Parson, Charles B	Pvt	K			
Pate, Alonzo L	Pvt	B			
Pate, Herman	Ck	E			
Pate, McIvey J	Pvt--1st Cl	C		W— Oct. 17-18	Nov. 26-18.
Patterson, Andrew	Pvt--1st Cl	Hq			
Patterson, Herbert W	Pvt	A	Dec. 6-18		
Patterson, Oscar M	Pvt--1st Cl	Hq			
Patterson, Robert F	Pvt--1st Cl	F		W— Sept. 29-18	Dec. 7-18.

Patterson, William D	Pvt	C		S— May 1-18	
Patton, Arthur C	Pvt	G			
Patton, James E	Sgt	D			
Patton, Tiry J. C	Pvt	M			
Patton, Joseph M	Cpl	H		W— Sept. 29-18 Deserted— Sept. 29-18	
Patton, Joseph W	Cpl	G		K— Sept. 29-18	
Patton, Sherman H	Pvt	M		K— Aug. 6-18	
Paulaskas, Joseph	Pvt	D			
Paulson, Manford P	Pvt	I			
Pavlak, John	Pvt	C		W— Oct. 17-18	Dec. 14-18.
Payne, George E	Pvt	F		W— Sept. 29-18	Dec. 17-18.
Pearce, Reuben	Pvt	G			
Pearce, Joseph L	Pvt	F		K— Jul 31-18	
Pearcy, Joe N	Cpl	H		W— Sept. 29-18	Dec. 19-18.
Pearson, Henry F	Pvt	H		W— Sept. 29-18	Dec. 19-18.
Pearson, Will Z	Pvt	I		K— Aug. 2-18	
Peavler, William C	Pvt	C	Dec. 6-18		
Peddicord, John H	Mus	Hq			
Pederson, Walter N	Pvt	MG		S— Sept. 27-18	Oct. 27-18.

Pederson, Arne	Pvt	D		K— Jul 18-18	
Peek, Lester	Pvt	A			
Peeler, Ellis E	Cpl	M			
Peeler, Lawrence T	Cpl	H		K— Sept. 29-18	
Peeples, Arthur W	Pvt	A			
Peeples, Frank E	Pvt	C	Dec. 6-18		
Peeples, James I	Pvt--1st Cl	MG		W— Sept. 27-18	Oct. 27-18.
Pelfry, Clement C	Pvt	G			
Pelfrey, Samuel J	Pvt	Hq			
Peltier, Rene	Pvt	A	Dec. 6-18	S— Mar 16-18	
Pennal, Robert A	Pvt	C	Dec. 6-18		
Pennell, Norman E	Pvt	A		W— Sept. 29-18 S— Mar 16-18	Nov. 11-18.
Pennell, Samuel P	Pvt	F		W— Sept. 2-18	
Pennington, Corley	Pvt--1st Cl	C			
Pergi, Frank	Pvt	B		W— Sept. 29-18	
Perkinson, Harry	Cpl	I		W— Sept. 29-18	Dec. 14-18.
Perrino, Dominick	Pvt	K	Dec. 6-18		
Perry, David T	Cpl	M		S— Oct. 31-18	Nov. 26-18.
Perry, John H	Pvt--1st Cl	K			

Perry, William J	Pvt	MG			
Perry, James R	Pvt	B		W— Sept. 29-18	
Perry, Seth E	Cpl	K		K— Sept. 29-18	
Perry, Thomas L	Pvt	I		K— Aug. 2-18.	
Person, Clyde C	Ck	F		W— Sept. 29-18	Nov. 20-18.
Peruzzini, Marreno	Pvt	K	Dec. 6-18		
Peterson, Harry C	Pvt	H	Dec. 6-18		
Peterson, Alphus R	Cpl	MG			
Peterson, Arthur M	Pvt--1st Cl	E		W— Sept. 29-18	Nov. 2-18.
Peterson, Harry E	Pvt	K			
Peterson, George L	Pvt	Hq		S— Aug. 11-18	
Petteway, Roger N	Pvt--1st Cl	A		T— Jan. 29-19	
Pettie, Edgar C	Pvt	E		W— Sept. 29-18	
Petty, Romeo	Pvt	I		K— Aug. 29-18	
Petrowich, Frank	Pvt	C			
Pettitt, George B	Pvt	C	Dec. 6-18		
Petty, Acy	Pvt	A			
Petty, John W	Ck	M			
Phelps, Lenwood	Bug	I		W— Oct. 10-18	Dec. 5-18.
Phelps, William J	Pvt--1st Cl	L		W— Oct. 17-18	

Phillips, Bennie L	Cpl	A			
Phillips, Clyde	Cpl	F		W— Oct. 10-18	Nov. 11-18.
Phillips, Don D	Pvt--1st Cl	MG			
Phillips, John S	Pvt	L		W— Oct. 10-18	Dec. 5-18.
Phillips, William J	Pvt	B			
Phillips, Amy B	Pvt--1st Cl	F		W— Sept. 29-18	
Phillips, Daniel	Pvt	F		T— Jul 24-18	
Phippen, Enos C	Pvt	L		W— Aug. 31-18	
Phipps, Charles	Pvt	Sup			
Piche, Andrew W	Pvt--1st Cl	H		W— Sept. 29-18	
Pickens, Emmerson J	Pvt	G			
Pieper, Albert H	Pvt	A			
Pierce, Ambrose F	Pvt	H		W— Sept. 29-18	
Pierce, Howard L	Col. Sgt	Hq		W— Sept. 29-19	
Pierce, Charles R	Sgt	G		K— Oct. 10-18	
Pierce, Luland W	Pvt--1st Cl	G		W— Oct. 17-18	
Pigett, Lindsey	Cpl	B		W— Sept. 29-18	
Pigford, Leon C	Pvt	H			
Pigg, Otto E	Sgt	Hq			
Pigue, Paul W	Cpl	K			

Pilgreen, John P	Pvt--1st Cl	D		W— Sept. 29-18	Dec. 8-18.
Pinkerton, LeRoy	Pvt	G		K— Oct. 18-18	
Pinkham, Lyman K	Pvt	Hq		W— Sept. 27-18	
Pipkin, William H	Pvt	C			
Pitisano, Antonio	Pvt	K	Dec. 6-18		
Pittman, Bryant A	Pvt--1st Cl	K			
Pittman, Jacob H	Sgt	G		W— Aug. 26-18	Jan. 4-19.
Pittman, John T	Sup. Sgt	K			
Pittman, William W	Pvt	Sup			
Pittman, Randolph	Pvt	D		K— Oct. 18-18	
Pittman, Robert L	Sgt	E		T— Sept. 10-18	
Pitts, Reuben A	Pvt	Sup		S— Dec. 8-18	
Plaskas, Joseph F	Pvt	MG		W— Sept. 29-18	Nov. 5-18.
Pleasant, Jasper R	Pvt	M			
Pleasant, William S	Pvt	I			
Pleasant, Major W	Pvt	H		W— Oct. 10-18	Dec. 2-18.
Plessinger, Richard	Pvt--1st Cl	M		S— Nov. 27-18	
Plonk, Marvin E	Pvt	H		W--Oct. 10-18	Dec. 2-18.
Plonk, Roy F	Cpl	A			
Plott, Guy A	Sgt	L		W— Oct. 17-18	Dec. 2-18.

Plummer, David C	Pvt	E		W— Sept. 29-18	
Pohlker, John A	Pvt	D		K— Oct. 14-18	
Poindexter, William F	Cpl	M		S— Oct. 31-18	Dec. 21-18.
Poland, John E	Pvt--1st Cl	Hq			
Poland, Samuel J	Pvt	G		W— Sept. 29-18	Nov. 11-18.
Polla, Ceasare	Pvt	L		W— Sept. 29-18	Dec. 2-18.
Pollard, Mack D	Pvt--1st Cl	I		S— Oct. 14-18	Nov. 26-18.
Polley, Chester	Corp	M		W— Oct. 10-18	Dec. 2-18.
Pollock, John C	Pvt--1st Cl	Hq		W— Sept. 29-18	Nov. 26-18.
Polly, Frank	Pvt	E		W— Oct. 17-18	
Poole, William L	1st Sgt	L			
Pope, Dallas A	Pvt--1st Cl	Hq			
Pope, Danie	Sgt	M		S— Oct. 14-18	Nov. 26-18.
Pope, Eldridge H	Ck	H			
Pope, John B., Jr	Pvt	A			
Pope, Oscar	Pvt--1st Cl	M			
Poppel, Henry	Pvt	E		K— Oct. 10-18	
Porcelli, Robert	Bglr	H		K— Jul 17-18	
Porter, Clarence R	Cpl	D		W— Sept. 29-18	Dec. 7-18.

Porter, Ralph J	Pvt--1st Cl	K			
Porter, Barney T	Pvt	C		S— Jan. 6-19	
Poston, Samuel McB	Pvt--1st Cl	F		W— Oct. 10-18	Dec. 22-18.
Poteat, George F	Pvt--1st Cl	C			
Potter, Robert A	Pvt--1st Cl	Hq			
Potter, Harry	Pvt	H		K— Sept. 29-18	
Potts, Ernest R	Pvt	F		W— Jul 24-18	Dec. 3-18.
Potts, Herman L	Pvt	I		W— Sept. 29-18	Dec. 5-18.
Potts, Thomas H	Pvt--1st Cl	L			
Powelka, Gilbert	Pvt	D		K— Sept. 29-18	
Powell, Houston E	Pvt	Hq			
Powell, Lester F	Cpl	Hq		W— Sept. 29-18	Nov. 11-18.
Powell, Frank	Pvt	Sup		W— Oct. 7-18	
Powell, Robert	Cpl	H		W— Sept. 29-18	
Prater, Guy C	Cpl	Hq		W— Sept. 29-18	Dec. 26-18.
Pratt, Robert L	Cpl	I		W— Aug. 31-18	Nov. 11-18.
Presley, James	Pvt	A			
Presley, Manson W	Pvt	Sup			
Pressley, Lonnie C	Pvt--1st Cl	MG			

Pressley, Dennie	Pvt	F		W— Oct. 17-18	
Pressley, Henry W	Pvt--1st Cl	M		W— Oct. 10-18	
Pressley, Jesse	Pvt	Sup		S— Dec. 9-18	
Pribianca, Alpine	Pvt	B		K— Sept. 27-18	
Price, Bo B	Pvt--1st Cl	B			
Price, James A	Pvt--1st Cl	H		W— Sept. 29-18	Nov. 26-18.
Price, Lane C	Ck	I			
Price, Malon A	Wag	Sup			
Price, Alphonso	Wag	Sup		S— Jan. 3-18	
Price, Grover C	Sgt	B		S— Sept. 20-18	
Price, Samuel G	Cpl	B		W— Sept. 29-18	
Price, Willard	Pvt	L		W— Aug. 2-18	
Prichard, James L	Pvt	K		W— Sept. 29-18	
Pritchard, Charlie A	Pvt	I			
Priddy, Arthur	Pvt	C			
Priddy, Maury W	Pvt	C	Dec. 6-18		
Pridgeon, Adolph	Cpl	E			
Prince, Herbert L	Cpl	I			
Pritchard, George L	Pvt	I			

Prosser, Roy W	Pvt	E		K— Sept. 29-18	
Pruett, Charles M	Pvt	C			
Pruitt, Fred C	B. Sgt	Hq			
Pruitt, John G	Pvt--1st Cl	Hq			
Pruitt, Rufus G	Cpl	H			
Pruitt, John C	Pvt--1st Cl	D		K— Sept. 29-18	
Pryer, Edward C	Bglr	B			
Pryor, William A	Pvt--1st Cl	G			
Pryor, Willie	Pvt	B			
Pryor, Cordel	Pvt--1st Cl	G		W & D— Sept. 29-18.	
Puckett, John A	Pvt--1st Cl	F		W— Sept. 29-18	Dec. 19-18.
Puckett, William L	Wag	Sup			
Pugh, William M	Pvt	C			
Pugh, William M	Sgt	Inf		T— Jul 17-18	
Pugh, William M	Sgt	D		W— Sept. 29-18 T— Feb. 4-18	Nov. 14-18.
Pulliam, Joshua	Pvt	K		S— Nov. 1-18	
Pully, Samuel C	Pvt	H			
Purcell, Douglas B	Pvt	L		T— Oct. 31-18	
Purifoy, Ray A	Pvt--1st Cl	D		W— Sept. 29-18	

Purnell, Edward D	Cpl	L		W— Oct. 17-18	Nov. 11-18.
Purvis, Dock W	Pvt	Hq			
Purvis, Whit C	Cpl	H		W— Sept. 29-18	
Quamn, Walter B	Pvt	A		M— Sept. 29-18	
Queen, Luther F	Pvt	K			
Quickstad, Arthur C	Pvt	H		W— Oct. 10-18	Dec. 19-18.
Quincy, Earnest L	Pvt	M			
Quinn, Harper	Pvt--1st Cl	E			
Quitzau, Peter Z	Cpl	D	Dec. 6-18		
Rabon, Herbert	Pvt	I			
Raby, John H	Pvt--1st Cl	F			
Rachley, Uda M	Sgt	H		S— Oct. 31-18	
Radcliff, Guy	Pvt	M			
Radford, Herman	Pvt	D			
Ragan, Emory E	Pvt	A			
Raines, James	Pvt	C	Dec. 6-18		
Rains, Simon G	Pvt	C		S— May 29-18	
Rainwater, Dink	Pvt	G		W— Sept. 29-18	
Rakers, John G	Pvt	K			

Ramer, Ben	Pvt	K		W— Sept. 29-18	Nov. 26-18.
Ramsuer, Daniel W	Pvt	K		S— June 19-18	
Ramsey, Charlie A	Pvt	K		W— Sept. 29-18	
Ramsey, Lee	Pvt	L		M— Aug. 31-18	
Randle, John C	Pvt--1st Cl	MG			
Randolph, J. B	Pvt	MG			
Rankin, Conway	Pvt	D	Nov. 2-18		
Rankin, John H	Ck	Sup			
Ranow, Clarence G	Pvt	E	Oct. 26-18		
Rasberry, Archie C	Cpl	F		W— Sept. 29-18	
Rash, Hershell H	Pvt	M		W— Sept. 29-18	Nov. 26-18.
Rash, Lee F	Cpl	H			
Rawls, Grady	Pvt--1st Cl	L		T— Feb. 8-19	
Rawls, Hampton	Pvt	MD			
Rawls, Herbert D	Cpl	Hq			
Rawls, Jud L	Pvt--1st Cl	H			
Rawls, Will D	Pvt	B		W— Oct. 10-18	Dec. 2-18.
Raxter, Ernest	Pvt	G		W— Sept. 29-18	
Raxter, Roy	Pvt	G		T— May 1-18	

Ray, Charles M	Pvt	C	Dec. 6-18		
Ray, Emory L	Pvt	B		W— Oct. 10-18	
Ray, Roy T	Pvt	Sup			
Raymond, Stanley A	Pvt	F		W— Oct. 10-18	Dec. 11-18.
Raynor, Lester C	Sgt	F			
Reader, George E	Pvt	A		K— Oct. 17-18	
Reardon, Michael J	Pvt	E		W— Sept. 29-18	Dec. 2-18.
Reaves, William J	Pvt	G			
Reavis, Raymond L	Pvt	G		W— Sept. 28-18	Dec. 1-18.
Rebaschio, Constantino	Pvt	K	Dec. 8-18		
Rector, John C	Pvt--1st Cl	MG			
Redman, Frank M	Pvt	H		T— Jul 6-18	
Redmond, William J	Pvt	M		S— Aug. 28-18	
Reece, Frank W	Pvt--1st Cl	F			
Reed, Floyd	Pvt	A	Dec. 6-18		
Reed, Jackson H	Pvt	Sup			
Reece, Grant	Pvt	F		W— Sept. 29-18	Dec. 18-18.
Reece, John W	Ck	D			
Reeves, Curtis E	Pvt	Sup			

Reeves, Ira E	Pvt	A		W— Aug. 25-18	Oct. 20-18.
Reeves, Ollie H	Cpl	L			
Reeves, Willie D	Pvt	A	Dec. 6-18		
Register, Max B	Sgt	H		W— Sept. 29-18	Dec. 20-18.
Reginall, John A	Cpl	E		W— Oct. 10-18	Dec. 7-18.
Reich, Pleasant A	Pvt	F			
Reid, Albert L	Pvt--1st Cl	MD		T— Jul 17-18	
Reid, George J	Pvt--1st Cl	C			
Reid, Henry F	Pvt	F		W— Jul 31-18	Nov. 26-18.
Reller, Frank H	Pvt	F		W— Jul 30-18	
Reneau, James F	Pvt--1st Cl	A			
Renegan, George D	Sgt	M			
Rentz, Henry L	Pvt	MD		T— Jan. 1-19	
Revan, John J	Pvt	I			
Reynolds, James H	Pvt	G		W— Sept. 28-18	Nov. 26-18.
Reynolds, Lemuel H	Pvt--1st Cl	K			
Reynolds, Louis S	Pvt--1st Cl	M			
Reynolds, Macon A	Pvt	MG			
Ray, Andrew B	Pvt	E		W— Sept. 29-18	Nov. 11-18.

Rhea, Eldridge	Pvt--1st Cl	E		W— Sept. 29-18	Oct. 29-18.
Rhinehart, Walter	Pvt	F			
Rhoads, Alex F	Pvt	H		K— Sept. 29-18	
Rhodes, Charlie	Pvt	A	Dec. 6-18		
Rhodes, Dallas	Cpl	M		W— Sept. 28-18	Dec. 5-18.
Rhodes, Gordon L	Sgt	F		K— Sept. 29-18	
Rhodes, Joseph	Pvt	L		W— Oct. 17-18	
Rhodes, Richard L	Mus	Hq		W— Sept. 2-18	
Rhodes, Walter J	Sgt	B			
Rhue, George N	Cpl	B		W— Oct. 10-18 T— Feb. 4-19	
Rhue, William C	Pvt	B		W— Oct. 17-18 T— Feb. 4-19	
Rice, Jeptha	Pvt--1st Cl	H		W— Sept. 29-18	Nov. 11-18.
Rice, Timothy L	Cpl	F		W— Sept. 29-18	
Rich, Carmel C	Pvt	A		K— Oct. 2-18	
Rich, William J	Pvt	B		W— Sept. 29-18	Oct. 27-18.
Richards, Homer	Cpl	D		T— Jul 22-18	
Richards, Zollie C	Pvt--1st Cl	Sup			
Richardson, Harvey	Pvt--1st Cl	B		K— Sept. 29-18	

Richardson, Ocee	Pvt	B		W— Oct. 10-18	Nov. 14-18.
Richardson, Curna	Pvt--1st Cl	A		W— Sept. 29-18	
Richardson, Ralph	Cpl	H		W— Sept. 29-18	Nov. 11-18.
Richards, Ernest L	Sgt	G			
Richard, John R	Cpl	G		W— Sept. 29-18	Dec. 14-18.
Rickard, William W	Pvt	Hq		S— Mar 15-19	
Riddick, Walter	Cpl	I			
Ridings, John T	Pvt	C			
Riggins, Will A	Mech	K		W— Sept. 29-18	
Riggs, Clem M	Pvt	K		K— Sept. 29-18	
Riggs, Earl H	Cpl	A			
Riley, John W	Cpl	A			
Riley, Lonnie T	Pvt--1st Cl	K			
Rinaldi, John	Pvt	K	Dec. 6-18		
Ritchie, Charlie F	Pvt				
Ritter, Eli M	Cpl	Hq			
Rivenbark, James A	Wag	Sup			
Rives, John S	Pvt	E		K— Sept. 29-18	
Roach, John G	Pvt	Inf			

Roache, Perry E. A	Pvt	C	Dec. 6-18		
Rone, Ellis B	Pvt	E		W— Oct. 10-18	Nov. 27-18.
Roark, Hendrick	Pvt--1st Cl	G		K— Oct. 10-18	
Roark, Dewey R	Sgt	D		W— Jul 18-18	Oct. 27-18.
Robbins, John, Jr	Pvt--1st Cl	D			
Robbins, Connie	Pvt--1st Cl	A			
Robbins, William E	Pvt	A		K— Sept. 29-18	
Roberson, Clenny	Pvt	E			
Roberson, James H	Pvt	L		W— Sept. 29-18	Dec. 5-18.
Roberts, Charles A	Pvt	A			
Roberts, Emery	Pvt	H		W— Aug. 20-18	Dec. 19-18.
Roberts, Sewell	Pvt	H		W— Sept. 29-18	Nov. 27-18.
Roberts, Harold P	Pvt	Hq		T— Oct. 10-18	
Roberts, Keifer W	Pvt	H		W— Sept. 29-18	
Roberts, Noah	Pvt	E		S— Sept. 10-18	
Roberts, Vance	Sgt	I			
Roberts, Wayne	Pvt	D			
Roberts, Willie	Sgt	M			
Robertson, Columbus W	Pvt--1st Cl	F		M— Oct. 18-18	Dec. 8-18.

Robertson, Harvey A	Pvt	H		S— Jul 6-18	
Robertson, Marvin C	Pvt	D		W— Oct. 10-18	Dec. 2-18.
Robinson, Cleburn	Pvt	E		S— Mar 16-19	
Robinson, Herman	Cpl	D		W— Sept. 29-18	Dec. 8-18.
Robinson, Stanley J	Pvt	I		T— Jan. 29-19	
Robinson, Thomas	Pvt	D		S— Oct. 1-18	Dec. 8-18.
Rochelle, Dumas W	Bug	MG		W— Sept. 29-18	Dec. 7-18.
Rochelle, James M	Cpl	Hq			
Rodgers, Bruce	Pvt	D		W— Aug. 20-18	
Rodgers, Glenn C	Pvt	B		W— Oct. 17-18	
Rodgers, Hardy	Pvt--1st Cl	G		D of W— Nov. 3-18	
Rodgers, Thomas B	Pvt	I		W— Sept. 3-18 W— Sept. 29-18	Dec. 2-18.
Roe, Vance	BSM	Hq		W— Sept. 29-18	
Roeckers, Henry B	Pvt--1st Cl	E		W— Oct. 10-18	Nov. 26-18.
Rogers, Claude H	Cpl	L			
Rogers, David J	Mus	Hq			
Rogers, Joe E	Pvt	I			
Rogers, Robert C	Pvt				
Rogers, Rufus M	Wag	Hq			

Rogers, Zollie C	Pvt	M			
Rogenburt, Frank	Pvt	L		W— Oct. 17-18	
Roland, Edward L	Pvt--1st Cl	Hq		S— Oct. 29-18	Dec. 1-18.
Roles, Nick P	Pvt	E		T— May 31-18	
Rollins, Griffin C	Pvt--1st Cl	H		S— Oct. 31-18	Dec. 2-18.
Romines, Arlie J	Pvt	L			
Romines, Frank	Pvt	L			
Roney, Lemuel L	Pvt--1st Cl	A			
Roper, Floyd	Cpl	D		W— Sept. 29-18	Oct. 27-18.
Rorie, James H	Pvt	Sup			
Rorie, John	Pvt--1st Cl	F		W— Sept. 29-18	Nov. 6-18.
Roscoe, Joseph E	Pvt--1st Cl	I		K— Aug. 31-18	
Roscoe, Decater F	Pvt	K		W— Oct. 10-18	
Roscoe, Guy P	Pvt--1st Cl	A			
Roscoe, Benjamin W	Pvt	H		W— Sept. 29-18	Nov. 16-18.
Rose, Chester A	Sgt				
Rose, Howard A	Pvt	C		W— Oct. 10-18 W— Sept. 1-18 W— Aug. 20-18	Dec. 12-18.
Rose, Norman A	Cpl	C		S— Oct. 1-18	

Rose, Elmer H	Pvt	H		W— Sept. 25-18 W— Sept. 29-18	
Ross, Albert C	Pvt	H			
Ross, Harry O	Pvt	I		T— Nov. 27-18	
Ross, Wendell A	Pvt	M		K— Aug. 4-18	
Ross, Emit C	Pvt	A			
Rosum, William	Pvt	D		T— Nov. 11-18	
Roten, Rufus F	Pvt--1st Cl	K			
Rountree, William L	Pvt	F		T— Oct. 28-18	
Rounsayell, William R	Pvt	H			
Rourk, Crawford L	Sgt	E			
Rowan, Thomes J	Sgt				
Rowell, Knox R	Cpl	E			
Rowlett, Barrett B	Pvt--1st Cl	L			
Royal, John H	Cpl	H			
Royal, Luby F	Sgt	C		T— Jul 22-18	
Royals, James M	Pvt--1st Cl	Hq		T— Mar 16-19	
Rudder, John H	Pvt--1st Cl	I			
Ruff, Andrew E	Pvt	K		K— Sept. 29-18	
Rumbley, Harrison W	Pvt--1st Cl	A			

Rummage, Fred A	Wag	Hq			
Ruple, Marion T	Sgt	MC			
Ruprecht, Albert	Pvt	H		K— Sept. 29-18	
Rusch, Ludwig H	Pvt	I			
Rush, Elmer J	Pvt--1st Cl	C			
Rushin, Grady E	Pvt--1st Cl	Hq		D of W— Jul 19-18	
Russ, Andrew M	Pvt	Sup			
Russ, Walter	Cpl			W— Sept. 28-18	
Rustin, Joel H	Pvt	A			
Rutledge, Henry T	Pvt	E		T— Jan. 22-19	
Rutledge, Jack D	Pvt	E		T— May 18	
Ryals, John P	Cpl	K		T— May 18-18	
Ryan, Ray B	Pvt	H		W— Sept. 29-18	
Rymer, John H	Pvt	A		W— Sept. 29-18	Dec. 19-18.
Sabalewski, Joseph A	Pvt	K	Dec. 6-18		
Sachrest, Luther F	Mech	Sup			
Sahms, Gus B	Pvt	M		S— Oct. 31-18	Nov. 26-18.
Salamas, Angelo	Pvt	L			
Sales, Thomas E	Pvt	I			
Salmon, David R	Pvt--1st Cl	C			

Salter, Benjamin R	Pvt	I		W— Aug. 25-18	
Salvers, Adelard	Pvt	K	Dec. 6-18		
Sammath, Lewis R	Sup Sgt	D			
Sams, Willie	Pvt	K		W— Oct. 10-18	Dec. 20-18.
Samuelson, Alfred M	Pvt	E		S— Dec. 3-18	Dec. 11-18.
Sandell, James L	Pvt	D	Dec. 6-18		
Sanderfer, Paul C	Sgt	M		W— Oct. 10-18	
Sanders, Albert L	Pvt	F			
Sanders, George	Sgt	D		W— Oct. 17-18	Nov. 6-18.
Sandlin, George S	Sgt	H		W— Aug. 20-18	Nov. 26-18.
Sandlin, John M	Pvt	F			
Sanford, Ralph N	Sgt	D		T— Jul 6-18	
Sanlin, Joe C	Pvt	D	Dec. 6-18		
Sansil, Bennie	Pvt	L		S— Sept. 15-18	Nov. 11-18.
Sasser, Henry	Wag	Sup		S— Nov. 11-18	
Sasser, Robert Y	1st Sgt	E		W— Sept. 29-18	
Santarossa, Antonio	Pvt	K	Dec. 6-18		
Sapp, Fred W	Pvt--1st Cl	MG		W— Oct. 17-18	Dec. 2-18.
Sartin, Lester W	Pvt	I			

Sasser, Roby	Cpl	E		W— Oct. 17-18	Dec. 3-18.
Sasser, Cordis L	Cpl	I			
Sasser, Dawson M	Pvt--1st Cl	F		W— Sept. 29-18	Nov. 6-18.
Sattler, John J	Pvt	G			
Sauls, John E	Sgt	C			
Saulsbury, Frank	Pvt	D			
Saunders, George W	Sgt	Hq		W— Sept. 29-18	Nov. 11-18.
Saunders, Coy L	Cpl	I		W— Aug. 5-18	
Saunders, Ozie T	Pvt	D		K— Oct. 10-18	
Saunders, Bowan D	Pvt	C		K— Oct. 17-18	
Savage, Claude S	Ck	A			
Sawyer, Herbert J	Bug	I			
Sawyer, John M	Cpl	K			
Sawyer, Lonnie C	Pvt	E		W— Sept. 29-18	
Sawyer, Miles F	Pvt	L		W— Sept. 29-18	
Saxerude, A. M	Mus	Hq			
Saxton, Thomas	Pvt	A		K— Sept. 29-18	
Sayre, Everett G	Pvt	A			
Scarboro, Eli P	Sgt	Sup			

Scardina, Anthony	Cpl	E		W— Sept. 29-18	Dec. 2-18.
Scarlett, Luther F	Pvt--1st Cl	G		W— Sept. 29-18	Dec. 2-18.
Scarpenter, Fred	Pvt	E			
Schaal, John H	Pvt--1st Cl	D			
Schaefer, Charles W	Pvt	K		K— Oct. 16-18	
Schemansky, Andrew	Pvt	K	Dec. 6-18	Deserted— Jan. 9-19	
Schatz, Charles C	Pvt	D			
Schatzer, William H	Pvt	K	Dec. 6-18		
Schepps, Louis	Pvt	K	Dec. 6-18		
Scher, Anthony J	Pvt	F			
Scherf, William H	Pvt	I		K— Oct. 10-18	
Schiller, Vincent F	Pvt	G		W— Sept. 29-18	
Schmitz, George W	Pvt	K		S— Oct. 14-18	
Schoen, Mike	Pvt	I			
Schofield, Phillip	Pvt	D	Dec. 6-18		
Schommer, Edward J	Pvt	M			
Schrieber, Charles J	Pvt--1st Cl	B		S— Nov. 27-18	Dec. 22-18.
Schroedel, John T	Pvt	B		W— Oct. 17-18	
Schroeder, Raymond F	Pvt--1st Cl	L			

Schroepfer, George C	Pvt	C			
Schultz, Frank A	Pvt	L		W— Oct. 17-18	Dec. 8-18.
Schumacher, Anthony C	Pvt--1st Cl	Hq		S— Oct. 1-18	Oct. 2-18.
Schunemann, Henry L	Pvt	H			
Schurke, Fred F	Pvt	F			
Schutrumph, Otto H	Cpl	A		W— Oct. 17-18 T— Feb. 1-19	Jan. 3-19.
Schwalenic, Paul	Pvt	A			
Schwam, Edward O	Cpl	O		W— Oct. 17-18	Oct. 17-18.
Schwartz, Edward	Pvt--1st Cl	G			
Schwartz, Elroy	Pvt	L		W— Oct. 10-18	
Schwellensattel, John H	Pvt	I		S— Jul 6-18	Nov. 26-18.
Schwer, Henry G	Pvt	B		W— Sept. 29-18	
Schwertferger, Otto G	Pvt	F		W— Sept. 29-18	Oct. 29-18.
Scott, Charlie D	Pvt	F			
Scott, Clarence V	Mech	E			
Scott, Claude O	Ck	K			
Scott, Clyde W	Pvt--1st Cl	E		W— Sept. 29-18	Dec. 2-18.
Scott, Curtis I	Pvt	E			
Scott, Hershall	Pvt	G		W— Sept. 29-18	Dec. 19-18.

Scott, Wilson H	Cpl	I			
Scott, Zeno Z	Pvt--1st Cl	MG			
Scott, Fletcher F	Cpl	Hq		W— Aug. 25-18	Sept. 20-18.
Scott, Leo	Pvt--1st Cl	H		W— Oct. 10-18 S— Mar 15-19	Feb. 5-19.
Scott, Robert E	Pvt--1st Cl	H		W— Sept. 25-18	
Scrabeck, Ollie S	Pvt	A		K— Sept. 29-18	
Seabolt, Edgar D	Sgt	D		S— Jul 6-18	
Seago, James L	Pvt--1st Cl	D		W— Sept. 29-18	Nov. 11-18.
Sears, James E	Pvt	I		W— Sept. 29-18	Dec. 2-18.
Sease, Claude N	Pvt	MG		S— Dec. 8-18	Jan. 19-19.
Seawell, John H	Cpl	F		W— Sept. 29-18	Dec. 30-18.
Seay, Ira D	Pvt	Hq	Dec 6-18		
Sebastian, Joseph B	Cpl	E		W— Jul 27-18	Oct 15-18.
Sebestian, William B	Pvt	E		W— Oct. 17-18	Dec. 3-18.
Seewald, Herman A	Pvt	H			
Seiffertt, Walter L	Mech	I		S— Nov. 11-18	Nov. 29-18.
Sekscinski, Antonio	Pvt	F			
Selby, Tucker M	Pvt	K			
Self, Joseph L	Pvt--1st Cl	F			

Sellers, George E	Pvt	Sup			
Sellers, Rufus E	Cpl	I		W— Sept. 29-18	Oct. 17-18.
Sellers, Thomas D	Pvt	L			
Sells, John	Pvt	I			
Sena, Roy V	Pvt	I			
Sepaugh, Columbus C	Pvt--1st Cl	Sup			
Sessoms, Henry A	Wag	Sup			
Setliff, Percy J	Pvt	I		S— Sept. 13-18	Nov. 2-18.
Sewell, Clarence V	Pvt	I			
Sewell, George A	Pvt--1st Cl	I			
Sforza, John	Pvt	K	Dec 6-18		
Shakelford, Will E	Pvt	A	Dec 6-18		
Shakelford, Jesse H	Sgt	MG		T— Nov. 6-18	
Sharp, Homer	Pvt	L		W— Oct. 10-18	Nov. 27-18.
Sharp, Love	Pvt--1st Cl	Hq			
Sharpless, Samuel B	Pvt--1st Cl	A		W— Oct. 17-18	Dec. 21-18.
Shaw, Cameron C	Pvt	I		W— Aug. 31-18	Dec. 31-18.
Shaw, Joseph J	Pvt	Hq		W— Sept. 29-18	Nov. 11-18.
Shaw, Josey D	1st Sgt	Sup			

Shaw, Lionel W	Pvt	Hq			
Shaw, Robert	Sgt	F			
Shaw, Duncan C	Cpl	L		K— Sept. 29-18	
Shawe, Joe F	Pvt	Hq			
Shearer, Walter C	Sgt	B			
Shetts, Wiley G	Pvt	K		W— Aug. 3-18	Nov. 14-18.
Shetts, William L	Pvt	K		K— Aug. 4-18	
Shell, Henry	B Sgt	Hq			
Shell, Marshall H	Cpl	MG			
Shelton, Oscar	Cpl	G		W— Sept. 29-18	
Shepard, Albert	Pvt	I		S— Nov. 27-18	
Shephard, Manuel A	Pvt	C		W— Sept. 29-18	Dec. 2-18.
Sheppard, Edwin F	Pvt	D		W— Jul 22-18	Dec. 2-18.
Sherdon, Herbert	Pvt	F		W— Oct. 10-18	Dec. 2-18.
Sherrill, Jack B	Pvt--1st Cl	G			
Sherrill, Raymond	Pvt	Hq			
Sherrill, William H	B S M	Hq			
Shillings, Marshall	Pvt	G		W— Sept. 29-18	
Shipley, Raymond S	Pvt	C		W— Aug. 31-18	Oct. 27-18.
Shock, Hoyle H	Pvt--1st Cl	MG		W— Sept. 29-18	

Shodall, Enoch	Pvt	F		W— Sept. 29-18	
Shoemake, Lester	Pvt	G			
Shoemaker, James F	Pvt	K		W— Aug. 3-18	Nov. 26-18.
Shoffner, James H	Sgt	C			
Shope, George	Pvt	H		T— Jul 12-18	
Shore, Charles E	Pvt	C			
Short, Clarence	Pvt--1st Cl	B			
Short, Frank R	Pvt--1st Cl	B			
Short, Marcus	Pvt	I		W— Aug. 31-18	
Short, William P	Sgt	D		T— Feb. 4-19	
Shows, Houston	Pvt	D	Dec 6-18		
Shrum, William M	Pvt--1st Cl	G			
Shuford, John E	Pvt	F			
Shultz, Edwin L	Pvt	Sup			
Sidberry, George W	Ck	K			
Sidney, George L	Pvt--1st Cl	G		K— Oct. 14-18	
Sidwell, John W	Pvt	M		K— Aug. 27-18	
Sigman, Ernest H	Cpl	C			
Sikes, Spurgeon	Pvt	A			

Siler, William V	Pvt	F			
Simcox, Henry F	Pvt	B		T— Feb. 4-19	
Simons, Frank	Pvt--1st Cl	F		W— Oct. 10-18	
Simpson, Thomas B	Pvt--1st Cl	F		W— Sept. 29-18	
Sinclair, Miles V	Cpl	Hq			
Sinclair, Johnnie W	Pvt	A		W— Sept. 29-18	
Sills, John C	Ck	C			
Sills, Tony	Pvt--1st Cl	L			
Silver, Elza M	Pvt	B			
Silver, Morris	Pvt	G			
Silverman, Sam	Pvt	L	Dec 6-18		
Simmerson, Walter G	Pvt--1st Cl	G		S— Oct. 20-18	Dec. 2-18.
Simmons, James S	Pvt--1st Cl	Ordn			
Simmons, Lawson	Cpl	I			
Simon, Peter J	Pvt	D	Dec 6-18		
Simpson, John S	Pvt--1st Cl	H			
Simpson, Stephen L	Pvt	Inf			
Sink, John A	Cpl	A			
Sink, John L	Pvt	H		W— Sept. 29-18	Dec. 7-18.

Sink, Jake	Pvt	Hq		W— Sept. 29-18	
Sinone, Guisseppe	Pvt	K	Dec. 6-18		
Skeens, Ed. G	Pvt	Ordn			
Skeeter, John W	Pvt--1st Cl	I			
Skidmore, Joseph B	Pvt	H		M— Sept. 29-18	
Skinner, Olin C	Pvt	D		W— Sept. 29-18	
Skipper, Everett J	Pvt	I			
Skipper, George B	Cpl	D			
Skyeveland, Frederick	Pvt	E		S— May 1-18	
Slagle, Alfred M	Cpl	G		K— Sept. 29-18	
Slaton, William R	Pvt	L	Dec. 6-18		
Slattery, James F	Pvt	K		K— Sept. 29-18	
Sloop, John B	Wag	Sup			
Sluder, Leet	Mech	H		S— Oct. 29-18	Nov. 20-18.
Small, James E	Pvt--1st Cl	M		S— Nov. 27-18	Dec. 3-18.
Small, Isom E	Cpl	I		S— Nov. 22-18	
Smiley, John D	Cpl	K		W— Sept. 29-18	Nov. 26-18.
Smith, Al V	Pvt--1st Cl	L			
Smith, Albert S	Pvt	F			
Smith, Albert L	Ck	F			

Smith, Alonza P	Pvt	B			
Smith, Arthur D	Pvt	L			
Smith, Arthur W	Pvt	B			
Smith, Avery M	Pvt	D		W— Oct. 17-18	Dec. 8-18.
Smith, Branzel	Pvt	B		W— Sept. 29-18	Dec. 2-18.
Smith, Carter H	Ck	D		W— Sept. 29-18	Dec. 7-18.
Smith, Charlie G	Pvt--1st Cl	C			
Smith, Clarence	Pvt--1st Cl	C			
Smith, Claude L	Pvt	H			
Smith, Curtis L	Cpl	G			
Smith, Daniel C	Pvt	D	Dec. 6-18		
Smith, Ed	Sup. Sgt	M			
Smith, Edgar B	Pvt	Hq			
Smith, Elzie	Pvt	H			
Smith, Francis H	Pvt	F			
Smith, Frank	Pvt--1st Cl	D	Dec. 6-18		
Smith, Fred	Cpl	B		W— Sept. 29 18	Dec. 2 18.
Smith, Fred O	Pvt	Hq		S— Oct. 29-18	Dec. 5-18.
Smith, George	Pvt	F			

Smith, Goodman	Pvt	A			
Smith, Harry B	Pvt	I			
Smith, James L	Pvt	G			
Smith, James P	Pvt--1st Cl	Hq		W— Aug. 30-18	Aug. 30-18.
Smith, James T	Pvt	Hq			
Smith, Jim	Pvt	F			
Smith, Jim	Cpl	B			
Smith, Joe M	Pvt--1st Cl	Hq			
Smith, John E	Pvt	Hq		W— Sept. 27-18	Nov. 11-18.
Smith, Joseph W	Pvt	D	Dec. 6-18	S— Dec. 24-18	Jan. 7-19.
Smith, Joseph W	Sgt	K			
Smith, Lyle W	Pvt	Inf			
Smith, Oliver L	Pvt	Sup			
Smith, Oscar E	Pvt	L	Dec. 6-18		
Smith, Otto A	Pvt	G			
Smith, Percy A	Pvt	H			
Smith, Ray S	Cpl	H			
Smith, Robert A	Pvt--1st Cl	K		S— Oct. 14-18	Dec. 2-18.
Smith, Robert L	Pvt--1st Cl	C			

Smith, Ronald J	Pvt	G			
Smith, Roscoe	Pvt	B		W— Sept. 29-18	Dec. 8-18.
Smith, Russell E	Sgt	D			
Smith, Thomas	Pvt	D	Dec. 6-18	S— Jan. 2-19	Jan. 9-19.
Smith, Tom	Pvt--1st Cl	B		W— Sept. 29-18	Dec. 7-18.
Smith, William C	Pvt--1st Cl	A		W— Oct. 17-18	Dec. 7-18.
Smith, Willie C	Cpl	E		W— Oct. 17-18	Nov. 26-18.
Smith, Willie B	Pvt--1st Cl	B			
Smith, Alex	Cpl	D		T— Jan. 20-19	
Smith, Banjamin	Pvt	A		W— Sept. 29-18	
Smith, Corbett	Pvt	B		S— Nov. 11-18	
Smith, Curtis E	1st Sgt	D		K— Jul 18-18	
Smith, Frank A	Cpl	L		W— Aug. 31-18	
Smith, George F	Pvt--1st Cl	B		W— Sept. 29-18	
Smith, Herbert W	Pvt	B		S— May 1-18	
Smith, John E	Pvt	C		S— Mar 15-18	
Smith, John I	Cpl	A		W— Oct. 17-18 T— Jan. 29-19	Dec. 2-18.
Smith, Mack B	Cpl	B		S— May 1-18	

Smith, Marvin B	Pvt	G		S— Oct. 20-18	
Smith, Rollie M	Pvt	B		W— Sept. 29-18	
Smith, Rufus	Pvt	MG		S— Nov. 14-18	
Smith, Thomas S	Pvt	C		S— May 1-18	
Smith, Walter M	Pvt	D		K— Sept. 29-18	
Smith, William C	Pvt--1st Cl	MG		K— Sept. 29-18	
Smith, William M	Pvt--1st Cl	L		T— June 22-18	
Smithwick, Sam	Cpl	B		W— Sept. 29-18	
Smoot, Arthur	Pvt	H		W— Oct. 10-18	Nov. 26-18.
Smuk, Joseph, Jr	Pvt--1st Cl	F			
Smuk, Matt	Pvt	F		W& D— Aug. 20-18--	Aug. 21-18.
Smyre, William A	Bug	D			
Sneed, Charlie M	Pvt	L			
Snelgrove, Cary C	Pvt	H		W— Sept. 29-18	Dec. 23-18.
Snoddy, John C	Sgt	L		W— Sept. 29-18	
Snow, Roy B	Pvt	D	Dec. 6-18		
Snow, Charles R	Pvt	G		K— Oct. 10-18	
Snyder, Chester E	Pvt	H		W— Sept. 29-18	Dec. 21-18.
Snyder, John C	Pvt	Inf			

Snyder, Roy B	Pvt	F		W— Sept. 29-18	Nov. 26-18.
Sobkowiak, John F	Pvt	I		S— Oct. 31-18	Dec. 2-18.
Sobolewski, Stanislav	Pvt	L	Dec. 6-18		
Sorrell, Rufus E	Sgt	H			
Sorrell, Arthur	Pvt	H		T— Feb. 4-19	
Sorte, Marvin	Sup. Sgt	L			
Soucek, Edward A	Pvt	A		W— Sept. 29-18	Nov. 8-18.
Southard, Dorsie G	Pvt	B			
Southards, Bert	Pvt--1st Cl	I			
Southwick, Robert L	Pvt--1st Cl	Hq			
Sowell, William H	Pvt	MG		S— Sept. 27-18	Oct. 10-18.
Spagnolio, Anton	Pvt	B			
Sparenberg, Anthony	Pvt	D	Dec. 6-18		
Spargo, Lon J	Pvt--1st Cl	Hq		W— Oct. 17-18	
Sparks, John C	Pvt	G		W— Sept. 28-18	Nov. 26-18.
Sparks, Lankford	Pvt--1st Cl	K			
Spear, George W	Pvt--1st Cl	C		K— Aug. 20-18	
Spears, Henry W	Pvt	E			
Spell, Henry C	Pvt	F		S— Mar 16-19.	
Spencer, Chester	Pvt	D			

Spencer, Clyde M	Cpl	C			
Spivey, Alonzo	Pvt	MG		W— Sept. 29-18	Nov. 8-18.
Spivey, Willie O	Pvt--1st Cl	F		W— Oct. 10-18	Dec. 14-18.
Spies, Herbert H	Pvt	C		W— Sept. 3-18	Nov. 10-18.
Spivey, William A	Pvt--1st Cl	C			
Sponeman, Louis	Cpl	I			
Spooner, Lawrence E	Pvt	G		W— Sept. 29-18	Dec. 16-18.
Spratt, John E	Pvt--1st Cl	C		W— Oct. 17-18	Dec. 7-18.
Springs, George	Sgt	F		W— Oct. 10-18	Oct. 27-18.
Sprinkle, Nathan D	Cpl	C			
Spruel, Charles W	Pvt	G		W— Sept. 29-18	
Spruill, Richard E	Pvt	I		W— Aug. 26-18	
Stacey, Beecher	Cpl	G		W— Sept. 29-18	Oct. 27-18.
Stack, Clyde W	Mech	M			
Stadsky, Jack	Pvt	L	Dec. 6-18		
Stafford, Fred A	Pvt	E		W— Sept. 29-18	
Staie, George J	Pvt	K			
Stainbeck, William A	Pvt	A		W— Sept. 29-18	
Stair, Elmer J	Pvt--1st Cl	L		W— Aug. 30-18	Aug. 30-18.
Stallings, Kenifer T	Pvt--1st Cl	F		W— Sept. 29-18	

Stallings, Paul	Pvt	G		K— Jul 26-18	
Stallsworth, Edgar L	Pvt--1st Cl	L		W— Sept. 29-18	
Stamey, Gerrard	Pvt	G		W— Sept. 29-18	Nov. 6-18.
Stampler, Elvis E	Pvt--1st Cl	L			
Stampson, Gas D	Pvt	H		W— Sept. 29-18	
Stanbury, Luther W	H. S. H. R	MG			
Stanford, Albert L	Pvt	K		K— Oct. 14-18	
Stanfield, Jerry B	Mech	H			
Stanley, Roy J	Pvt	D		W— Aug. 21-18	Nov. 1-18.
Stanley, Harvey C	Cpl	C		S— May 1-18	
Stanton, Claude	Pvt--1st Cl	C		S— Oct. 28-18	
Starba, Stanislaw	Pvt	L	Dec. 6-18		
Starling, Willie D	Pvt--1st Cl	A			
Starling, James L	Pvt	F		W— Oct. 17-18	
Starnes, Mack W	Pvt--1st Cl	MG			
Staton, Harry	Cpl	F			
Stavos, Benoni J	Pvt	E		W— Sept. 29-18	
Steed, Hubert F	Sgt	L			
Steele, Finch	Cpl	E			

Steele, James F	Pvt	B		W— Oct. 17-18	Nov. 27-18.
Steelman, Thomas S	Cpl	E		W— Sept. 29-18	Oct. 29-18.
Steen, Edward P	Pvt	L			
Steinard, Edward H	Pvt	C		W— Aug. 31-18	Aug. 31-18.
Stephens, Clarence W	Pvt--1st Cl	E		W— Sept. 29-18	Dec. 14-18.
Stephens, Thomas W	Pvt	D		W— Sept. 29-18	
Stephenson, Cornelius W	Sgt	M			
Stephenson, Hallie	Pvt--1st Cl	K			
Stephenson, John	Ck	G			
Stephenson, Lucian H	Pvt	D		W—O ct. 17-18	Dec. 14-18.
Stephenson, William B	Pvt	H		K— Sept. 29-18	
Stephenson, Chester L	Sgt	C		W— Sept. 29-18 S— Mar 15-19	Oct. 27-18.
Steelman, James D	Pvt--1st Cl	E		K— Sept. 29-18	
Stern, Julian	Pvt	Hq			
Sterrenburg, Charles	Pvt--1st Cl	B			
Stevens, Greer C	Pvt--1st Cl	H			
Stevens, Leamon	Pvt	C		W— Oct. 17-18	Oct. 17-18.
Stevens, Stitcal	Bug	G		W— Oct. 17-18	Dec. 2-18.
Stevenson, Thomas F	Pvt	D	Dec. 6-18	T— Jan. 29-19	

Stewart, Fred E	Cpl	A			
Stewart, Arthur R	Pvt	A		W— Sept. 29-18	
Stillman, J. Leslie	Pvt	G		K— Jul 26-18	
Stinebaugh, Lloyd	Pvt	I			
Stinson, Johnnie W	Pvt	L	Dec. 6-18		
Stock, Henry	Pvt--1st Cl	B		W— Sept. 29-18 T— Feb. 4-19	Nov. 11-18.
Stockdale, Lee J	Pvt	A		W— Sept. 29-18	Dec. 14-18.
Stoeckel, Vincent H	Pvt	I			
Stokes, Grover C	Sgt	B		W— Sept. 29-18	Dec. 2-18.
Stokes, Ray	Pvt	B		S— Mar 15-19	
Stone, Jack B	Pvt	B			
Stone, Mitchell	Cpl	I		W— Oct. 17-18	Dec. 5-18.
Stone, Richard S	Bug	A		S— Oct. 14-18	
Stokes, Jacob B	Pvt	A		W— Sept. 29-18	
Stone, Robert H	Pvt	H		T— Jul 6-18	
Stoner, Enos A	Pvt	G			
Stonefield, Albert C	Pvt	K		W— Sept. 29-18	
Stoner, Lea R	Pvt	G		S— Oct. 16-18	

Storie, James R	Cpl	B		W— Oct. 17-18 T— Feb. 4-19	Dec. 2-18.
Stover, James M	Pvt	B			
Strandberg, Daniel	Pvt	K		W— Sept. 29-18	
Strauss, Frank B	Sgt	H		W— Sept. 29-18	Nov. 20-18.
Strickland, Clyde	Pvt--1st Cl	G		W— Oct. 10-18	Dec. 23-18.
Strickland, John H	Pvt--1st Cl	K			
Strickland, Leroy	Cpl	M		W— Sept. 29-18	Nov. 11-18.
Strickland, Percy L	Pvt	Hq			
Strickland, Wadus	Pvt	M			
Strickland, Walter A	Cpl	A			
Strickland, Claude	Cpl	M		W— Sept. 29-18	
Strobel, Ed	Pvt	L		W— Sept. 17-18	
Strohbusch, Richard L	Pvt	I	Dec. 6-18		
Strom, Frederick	Pvt	K			
Stromquist, Eric	Pvt	Hq			
Strong, John W	Pvt	G			
Stroup, Carlos C	Pvt--1st Cl	G		W— Oct. 17-18	
Stuart, Charles B	Pvt	D			
Stubbs, William J	Mess Sgt	Sup			

Stuempler, Herman	Pvt	I			
Stultz, Rex B	Pvt	F		W— Sept. 29-18	
Stumer, William	Pvt	F			
Sturgill, Frank J	Pvt	H			
Styles, Andy	Pvt	G		K— Sept. 29-18	
Sugg, Travis	Pvt--1st Cl	D			
Sugg, Arsene F	Pvt	MG		S— Nov. 1-18	
Suggs, Curtis C	Pvt	MG			
Suggs, Joseph W	Cpl	A			
Sullivan, George L	Pvt--1st-Cl	Hq		W— Sept. 29-18	Jan. 18-19.
Sullivan, Homer	Pvt--1st-Cl	E			
Sullivan, Obie T	Pvt	Hq			
Summerlin, Clarence	Pvt	E		W— Sept. 29-18	Dec. 2-18.
Summerlin, Lee	Pvt--1st-Cl	E			
Summers, Joseph D	Cpl	H			
Sumner, Lemuel H	Pvt--1st-Cl	M			
Sumner, Preston	Sgt	H		W— Oct. 10-18	
Sumner, Wiley	Pvt--1st-Cl	A		W— Sept. 3-18	
Sumrell, Eugene	Sgt	D		W— Oct. 17-18	

Sunski, Joe	Pvt--1st-Cl	B			
Sunblad, Anton W	Pvt	Hq			
Sundin, Theodore	Pvt--1st-Cl	L		W— Oct. 10-18 T— Feb. 8-19	Dec. 20-18.
Surles, Munroe	Cpl	C			
Surratt, William P	Pvt	G		K— Oct. 10-18	
Sussman, Ellis	Cpl	H		W— Aug. 26-18	Nov. 1-18.
Suttles, William R	Pvt	A		M— Oct. 10-18	
Sutton, Robert	Pvt--1st-Cl	B		K— Sept. 29-18	
Svarvari, Edwin	Pvt	Hq	Dec. 6-18		
Svenson, Carl	Pvt	M		T— Feb. 4-19	
Swain, Edison	Cpl	A			
Swart, Herbert E	Pvt	C		S— Oct. 1-18	
Sween, Herbert J	Pvt	A		W— Oct. 17-18	
Sween, Luddy	Pvt	A		S— Nov. 1-18	
Sweet, Orin	Pvt	MG		K— Sept. 29-18	
Sweet, Raymond C	Pvt	E		K— Sept. 29-18	
Swicegoode, Grover C	Cpl	H			
Swim, Robert L	Pvt--1st-Cl	D	Dec. 6-18		
Swindell, George W	Pvt	MG		W— Sept. 29-18	Oct. 27-18.

Swinson, John C	Sgt	K			
Swomley, William H	Pvt	B		T— Feb. 4-19	
Sykes, John E	Bug	G		S— Oct. 28-18	Nov. 11-18.
Sylvester, Howard M	Pvt	D	Dec. 6-18		
Tabbert, Rudolph	Pvt	D	Dec. 6-18		
Tally, Fred W	Mech	G			
Tally, Thomas A	Cpl	E			
Tally, William H	Pvt--1st-Cl	G		W— Sept. 29-18	
Tankersley, Henry	Pvt	Hq			
Tanner, Ernest B	Pvt--1st-Cl	Hq		T— Sept. 13-18	
Tate, Charlie G	Pvt--1st-Cl	D		W— Sept. 29-18	Nov. 1-18.
Taylor, Carl	Sgt. Bug	Hq			
Taylor, Christopher C	Pvt	M		W— Sept. 1-18	Oct. 27-18.
Taylor, Fred A	Pvt--1st-Cl	Hq			
Taylor, Irby B	Pvt	D	Dec. 6-18		
Taylor, King D	Pvt	F		W— Aug. 25-18 W— Sept. 29-18	Dec. 2-18.
Taylor, Pearlie J	Pvt	H		W— Sept. 29-18	Dec. 19-18.
Taylor, Samuel J	Pvt	A			
Taylor, Thomas L	Sgt	MG			

Taylor, Timmie D	Cpl	G		W— Sept. 29-18	Nov. 16-18.
Taylor, William H	Wag	Sup			
Taylor, Willie B	Pvt	B		W— Sept. 29-18	Dec. 2-18.
Tayloe, Johnathan H	Pvt--1st-Cl	I		K— Aug. 8-18	
Taylor, Carrie	Pvt	D		W— Jul 24-18	
Taylor, William	Pvt	F		K— Sept. 29-18	
Taylor, William	Pvt--1st-Cl	D		W— Sept. 29-18	
Taylor, William Z	Pvt--1st-Cl	D		W— Oct. 17-18	
Teague, John F	Pvt--1st-Cl	C			
Tedder, Rush W	Cpl	Hq			
Tedder, Sam C	Pvt	I			
Tetters, George W	Cpl	A			
Tenny, John W., Jr	Sgt	I			
Temple, Oscar	Cpl	M		T— Feb. 4-19	
Templeton, Frederick	Pvt	E		S— Oct. 29-18	
Terry, Avery C	Pvt	K		K— Aug. 26-18	
Terry, Claude P	Pvt	C		W— Oct. 17-18	
Terry, Arthur	Pvt	B			
Terry, John C	Pvt	MG			
Terry, Joseph	Pvt	K			

Tew, Lonnie	Sgt	F		W— Sept. 29-18	Dec. 19-18.
Tew, Columbus B	Pvt	H		W— Sept. 25-18	
Thaggard, Wayman T	Pvt	H		K— Sept. 29-18	
Tiemann, Walter H	Pvt	A		K— Jul 22-18	
Thigpen, Alpha	Pvt--1st-Cl	I		K— Aug. 8-18	
Thigpen, John	Pvt	D			
Thigpen, Robert W	Pvt	F			
Thigpen, Thomas L	Sgt	H		W— Sept. 29-18	Oct. 27-18.
Thoemke, George G	Cpl	L			
Thomas, Ervin E	Pvt	I			
Thomas, Francis	Pvt	I			
Thomas, George E	Mus	Hq			
Thomas, Harry L	Pvt	Ord			
Thomas, James	Cpl	MG			
Thomas, William C	Sgt	D			
Thomas, Edgar G	Pvt	A		S— Aug. 31-18	
Thomas, Frenk L	Cpl	F		W— Sept. 29-18	
Thomas, Horace G	Sgt	D		T— Aug. 31-18	
Thomas, Jesse	Pvt	D		T— May 4-18	

Thomas, Lonnie J	Pvt	G		K— Sept. 29-18	
Thompson, Clarence	Pvt	Hq		W— Sept. 29-18	Dec. 2-18.
Thompson, Clarence B	Pvt--1st-Cl	D		W— Oct. 10-18	Dec. 2-18.
Thompson, Ernest H	Pvt	H			
Thompson, Frank Y	Cpl	F		W— Sept. 29-18	Oct. 27-18.
Thompson, George A	Pvt	M		W— Sept. 29-18	Dec. 2-18.
Thompson, George G	Pvt--1st-Cl	E			
Thompson, Henry T	Wag	Sup			
Thompson, Hugh A	Sgt	Hq			
Thompson, Robert C	Pvt	B			
Thompson, Zack	Pvt--1st-Cl	G			
Thomason, Solomon L	Pvt	C		K— Oct. 17-18	
Thompson, Albert M	Pvt	A		T— Feb. 4-19	
Thompson, Clarence D	Pvt	D		T— Jan. 13-19	
Thompson, Joe W	Wag	Sup		W— Sept. 4-18	
Thompson, John W., Jr	Cpl	E		W- -Sept. 29-18	
Thompson, Leavy	Pvt	L		S— Oct. 1-18	
Thompson, Nathaniel B	Pvt	I		W— Oct. 10-18	
Thompson, Robert L	Cpl	D		W— Sept. 29-18 T— Jan. 20-19	Dec. 14-18.

Thompson, Travis	Pvt	D		K— Jul 22-18	
Thompson, Warren D	Pvt	E		S— June ?-18	
Thorne, Jesse	Pvt	C		T— June ?-18	
Thorne, Thomas J	Pvt	C		T— June ?-18	
Thorne, James O	Pvt	Hq			
Thorne, William M	Pvt	A			
Thrasher, Algernon E	Pvt	Hq	Dec. 6-18		
Thrasher, James A	Cpl	B		W— Sept. 26-18	
Thurman, Erasmus	Ck	K		W— Sept. 29-18	
Thurman, Litton T	Sgt	H			
Tilly, Austin W	Pvt	F		K— Oct. 10-18	
Tilly, George W	Pvt--1st-Cl	K		W— Sept. 29-18	
Tilly, Thomas	Cpl	L		W— Oct. 17-18	
Tilghman, Albert	Pvt	I			
Tillert, Robert F	Pvt	MG		W— Oct. 10-18	Dec. 2-18.
Tillman, Martin L	Wag	Sup			
Tilton, Edward L	Cpl	E		W— Sept. 29-18	
Tims, William C	Sgt	C			
Tiney, Eason	Pvt	I		K— Aug. 8-18	

Todd, August	Pvt	A		W— Sept. 29-18	
Tolar, Empie R	Cpl	G			
Tolefson, Auton G	Pvt	K			
Toller, Jesse L	Cpl	F		W— Sept. 29-18	Dec. 14-18.
Tollett, Carlisle S	Sgt	G			
Tomblin, Whiteford	Pvt	F		W— Sept. 29-18	Dec. 29-18.
Tomkins, George B	Pvt--1st- Cl	G			
Tomlinson, James M	Ord. Sg	Ord			
Tomlinson, Robert W	Sgt	F			
Tonelis, Frank	Pvt	L		W— Aug. 31-18	
Townsend, Claude C	Pvt	G			
Towrn, Wiley S	Pvt	E		K— Sept. 29-18	
Treadway, Troy	Pvt	G		W— Sept. 3-18	
Trinkle, Clyde C	Mech	E			
Triplett, Ralph H. P	Pvt	D		W— Sept. 29-18	Dec. 7-18.
Triplett, John O	Pvt--1st- Cl	I			
Triplett, Lark L	Pvt	A		S— Mar 16-19	
Triplette, Crattie	Pvt	C		T— Jul 6-18	
Trott, Eli G	Pvt	D		W— Sept. 29-18	

Troy, Ernest L	Pvt	H		S— Aug. 2-18	
True, Willis M	Cpl	L		W— Sept. 29-18	Dec. 23-18.
Truitt Ralph	Pvt	Hq	Dec. 6-18		
Trull, Ed. F	Cpl	A		W— Sept. 29-18	Dec. 2-18.
Tuck, Roscoe	Pvt--1st Cl	Hq			
Tucker, Charlie M	Pvt	D			
Tucker, William B	Pvt	E		W— Sept. 29-18	Dec. 7-18.
Tucker, Harley L	Cpl	Hq		K— Jul 17-18	
Tue, James E	Cpl	L		S— Oct. 28-18	
Tue, John T	Mech	L		S— Oct. 30-18	
Tullis, Julian P	Pvt	L	Dec. 6-18		
Turlington, Jesse E	Pvt--1st Cl	Md		T— Jul 17-18	
Turlington, Jesse T	HS	Sup			
Turner, David D	Pvt	K		K— Sept. 29-18	
Turner, Charles E	Pvt	MD			
Turner, Lyles G	Pvt	C		S— May 1-18	
Turner, China	Pvt--1st Cl	D		W— Jul 18-18 W— Sept. 29 18	Dec. 2-18.
Turner, Fuller C	Pvt	A			
Turner, James W	Pvt	D		W— Oct. 10-18	Dec. 8-18.

Turney, Herbert L	Pvt	A		W— Sept. 29-18	
Turrentine, Herbert S	Pvt	C		K— Aug. 31-18	
Tuttle, Troy W	Pvt--1st Cl	F		W— Sept. 29-18	Dec. 14-18.
Tweeten, Oliver D	Pvt	E		T— Jul 12-18	
Twiford, Elwood B	Pvt--1st Cl	A		W— Sept. 29-18	
Twine, Dempsey J	Cpl	I			
Tyndall, George D	Pvt--1st Cl	D			
Tyndall, Nick T	Pvt	B		T— Feb. 14-19	
Ullom, Harold	Pvt	L	Dec. 6-18		
Ulrich, Henry W	Pvt	G		W— Sept. 29-18	Dec. 19-18.
Umstead, Irvin E	Pvt	MG			
Umphlette, Claude L	Pvt--1st Cl	D		K— Oct. 10-18	
Underwood, Edwin A	Pvt	K		W— Oct. 10-18	Nov. 26-18.
Underwood, Ernest L	Pvt	Hq		W— Aug. 19-18	
Uzzell, Nayard	Pvt--1st Cl	K			
Vachon, Wilfred	Mech	Sup		T— Jan. 20-19	
Vaitz, Steve	Pvt	A		W— Oct. 17-18	Dec. 1-18.
Valentine, Marcel G	Pvt	K		W — Sept. 29-18 D— Sept. 30-18	
Velerio, John	Pvt--1st Cl	MD	Oct. 28-18		

Vandever, Virgil E	Cpl	G			
Van Dyke, Edgar L	Pvt	K			
Vanhorn, John W	Pvt	C			
VanRaden, Ralph	Pvt	Hq			
Varner, Andrew H	Pvt--1st Cl	D			
Vasser, Joseph W	Pvt	L	Dec. 5-18		
Vassillon, Alexander	Pvt	Hq	Dec. 5-18		
Vaughn, Joe H	Sgt	E			
Vaughn, Alton C	Pvt--1st Cl	B			
Vaughn, Charlie J	Pvt	H			
Vaughn, Elbert L	Pvt	B		W— Sept. 29-18	Sept. 29-18.
Vaughn, Ernest G	Pvt--1st Cl	MD	Oct. 28-18		
Vaughn, George	Pvt	A		W— Sept. 29-18	Nov. 11-18.
Vaughn, Dock T	Pvt	B		K— Oct. 18-18	
Vaughn, Hardy	Pvt	B		T— Feb. 14-19	
Vaughn, Henry H	Pvt	B			
Vaughn, Sam T	Pvt	I		W— Sept. 29-18	Dec. 5-18.
Vawter, Skyler B	Pvt	C		W— Sept. 29-18	
Verderosa, James	Pvt	L	Dec. 5-18		

Verderosa, Joseph C	Pvt	L	Dec. 5-18		
Vereen, Joseph J	Cpl	D		W— Sept. 29-18	Dec. 8-18.
Verna, Atillio	Pvt	A	Dec. 5-18		
Verzwyvert, John A	Pvt	L	Dec. 5-18		
Vester, James B	Pvt--1st Cl	B		W— Sept. 29-18	
Vetto, Loney	Pvt	L	Dec. 5-18		
Vezolles, Harold A	Bug	F		S— Aug. 13-18	Nov. 11-18.
Viar, Willie B	Pvt--1st Cl	K			
Viebrock, Charles	Pvt	L	Dec. 5-18		
Vigdorowitz, Max J	Pvt	L	Dec. 5-18		
Vines, Charles C	Sgt	A			
Vissichelli, John	Pvt	L	Dec. 5-18		
Vitiello, Rocco	Pvt	L	Dec. 5-18		
Vogel, Theodore K., Jr	Sgt	G		Discharged— Jul 16-18	
Voigt, Edward	Pvt	L	Dec. 5-18		
Voigt, Henry	Pvt	L			
Volpe, Dominick	Pvt	Hq	Dec. 5-18		
Voogd, Abe O	Sgt	F			
Vowell, Edward D	Pvt	L		W— Sept. 29-18	Dec. 21-18.

Vuichard, Will	Pvt	A			
Vandiver, Earl E	Pvt	M			
Wachsmuth, Harry L	Pvt	C			
Wachtler, Joseph M	Pvt	I		W— Oct. 10-18	Dec. 8-18.
Wade, Charles A	Pvt	G		W— Aug. 26-18	
Wadley, Carl L	Pvt	D		S— Dec. 24-18	
Wafflard, Lender	Pvt	A			
Waggoner, Bryan L	Bug	L		W— Sept. 29-18	
Wagner, Albert J	Pvt	Sup			
Wagner, Arnold L	Sgt	Inf		T— Nov. 6-18	
Wagner, Irvin B	Sgt	I		S— Aug. 9-18	Dec. 12-18.
Wagooner, Frank S	Pvt	E			
Wainwright, Everett E	Pvt	L	Dec. 6-18		
Wate, John E	Pvt--1st Cl	L		T— Feb. 8-18	
Waldrop, C. J	Pvt	Sup			
Walker, Aaron	Pvt--1st Cl	F			
Walker, Amos	Pvt	F		D— Aug. 25-18	
Walker, Carroll P	1st Sgt	D		W— Aug. 7-18 T— Jul 22-18	
Walker, Colonel R	Pvt	C		W— Aug. 31 18	

Walker, Clifford E	Sgt	C		T— Oct. 1-18	Jan. 8-19.
Walker, Debrial G	Pvt--1st Cl	G		W— Sept. 29-18	Nov. 11-18.
Walker, Edgar	Pvt	F		W— Sept. 29-18	
Walker, Harvey	Pvt	G		W— Sept. 29-18	
Walker, John W	Pvt	G		W— Oct. 10-18	Dec. 24-18.
Walker, John W	Pvt	H	Nov. 27-18		
Walker, Louis W	Pvt	G			
Walker, Mann	Cpl	K		W— Sept. 29-18	Nov. 20-18.
Walker, Otis	Pvt	D	Dec. 6-18		
Walker, Robie E	Pvt	H		W— Oct. 10-18	Dec. 19-18.
Walker, Roscoe B	Pvt	I		W— Oct. 10-18	
Walker, Samuel P	Pvt	I			
Walker, Urius I	Cpl	G			
Walker, William A	Pvt	D		W— Sept. 29-18	
Wall, Jesse D	Pvt	H		W— Oct. 10-18	
Wall, Ira E	Pvt	E	Dec. 12-18		
Wall, John L	Pvt	I		W— Oct. 10-18	Dec. 14-18.
Wall, Johnnie	Pvt	A			
Wall, Everett E	Pvt--1st Cl	Inf		T— Jul 5-18	

Wall, Trossie G	Wag	Sup		S— Nov. 11-18	
Wallace, David N	Sad	Sup			
Wallace, Eugene	Pvt--1st Cl	L		T— Feb. 8-19	
Wallace, Fred E	Pvt	F		W— Oct. 17-18	
Wallace, Theodore B	Cpl	L			
Wallace, Virgil	Pvt	L			
Wallace, William V	Pvt	D		W— Oct. 10-18	
Wallach, Abraham	Pvt	L	Dec. 6-18	T— Feb. 8-18	
Wallett, Wesley W	Pvt	L	Dec. 6-18		
Walls, Frank S	Pvt--1st Cl	D		S— Sept. 15-18	Nov. 26-18.
Walter, David C	Pvt	F	Oct. 26-18		
Walters, William J	Pvt	I	Dec. 6-18		
Walton, David J	Pvt	G			
Wamser, Fred	Pvt	K		W— Sept. 29-18	Dec. 5-18.
Ward, Edward J	Cpl	H			
Ward, George B	Pvt	M			
Ward, John G	Pvt--1st Cl	Hq		W— Sept. 29-18	
Walston, Britton	Pvt	C		T— June 1-18	
Walston, Leslie	Pvt	K		W & D— Aug 4-18.	

Walston, Willie	Pvt	D		K— Sept. 29-18	
Wanninger, Leo	Pvt	L		W— Sept. 29-18	
Ward, Ervin	Sgt	B		Discharged— Jul 20-18.	
Ward, George D	Pvt	D		W— Sept. 29-18	
Ward, John E	Pvt	F		T— June 1-18	
Ward, Robert L	Mech	Sup			
Ward, William L	Pvt--1st Cl	C		W— Oct. 17-18	Nov. 26-18.
Warner, Clyde A	Pvt	Hq			
Warren, Charlie A	Pvt--1st Cl	K			
Warren, James K	Pvt	M		W— Oct. 6-18	
Warren, John L	Pvt--1st Cl	E			
Warren, Joseph I	Pvt	M			
Warren, Louis W	Corp	D		W— Oct. 17-18	Dec. 2-18.
Warren, Samuel J	Pvt--1st Cl	H			
Washburn, Jeff	Pvt	Hq		W— Sept. 29-18	
Waterfield, Lloyd L	Pvt	L		K— Jul 31-18	
Waters, Thomas P	Pvt	Sup		T— Feb. 22-18	
Watkins, Claude	Pvt	I			
Watkins, Clyde E	Corp	F		K— Sept. 29-18	

Watkins, Ernest E	Pvt	A			
Watkins, Francis M	Pvt	F	Oct. 26-18		
Watkins, Walter D	Pvt	D		W— Sept. 29-18	Dec. 22-18.
Watson, Claude W	Pvt--1st Cl	G		W— Oct. 10-18	
Watson, Herbert M	Pvt--1st Cl	MG			
Watson, Joseph C	Sgt	L			
Watson, Millard P	Pvt--1st Cl	E			
Watson, Samuel E	Pvt	D		W— Oct. 17-18	
Watson, William J	Pvt	D	Dec. 6-18		
Watson, William J	Pvt	L		W— Sept. 29-18	Oct. 27-18.
Watts, Carl D	Corp	M			
Watts, Claude A	Pvt	C		T— Oct. 17-18	
Way, Alfred A	Pvt	A		W— Oct. 17-18	Nov. 26-18.
Wayne, Harvey C	Pvt--1st Cl	D		K— Jul 21-18	
Weakly, Charles T	Corp	C		T— Jul 13-18	
Weaver, Charlie T	Pvt	D		W— Sept. 29-18	
Weaver, George A	Corp	Hq		S— Nov. 26-18	
Weaver, Henry G	Pvt	G			
Weaver, Hubert O	Corp	A			

Weaver, Oseo	Pvt	G			
Weaver, Walter E	Pvt	H		T— Jul 12-18	
Weaver, Wayne A	Pvt--1st Cl	H		W— Sept. 29-18	Nov. 11-18.
Weaver, William A	Pvt--1st Cl	Hq		S— Mar 16-19	
Weaver, Wiley B	Corp	B			
Weaver, William A	Pvt	D			
Webb, Berry C	Pvt	D	Dec. 6-18		
Webb, Fred L	Pvt	L		W— Aug. 7-18	
Webb, Herbert S	Pvt--1st Cl	D		S— Sept. 28-18	
Webb, Percefur L	Pvt	E			
Webb, Ernest W	Sgt	K			
Webb, Vernon F	Pvt	D		K— Sept. 29-18	
Webster, George D	Pvt--1st Cl	F		W— Sept. 29-18	
Webster, Lonney A	Corp	F		W— Sept. 29-18	Dec. 15-18.
Weed, Cary J	Pvt	M	Dec. 6-18		
Welch, Fred	Pvt	D		K— Sept. 29-18	
Wegescheide, George S	Pvt	K			
Wells, James L	Pvt	D		T— Aug. 25-18	
Welbourn, George B	Pvt	Ord			

Welch, Andrew F	Pvt	F			
Welch, James D	Pvt	MG			
Wells, Vernie F	Sgt	G		T— Oct. 29-18	
Wellman, Leo V	Pvt	MG			
Wells, George D	Pvt--1st Cl	G			
Wells, Harry E	Pvt	Hq			
Wells, John I	Pvt	Hq			
Wells, John T	Sgt	E			
Wells, Lawson J	Wag	Sup			
Wells, Raymond E	Wag	Sup			
Wells, Wiley E	Pvt	Ord			
Wemsier, Peter G	Pvt--1st Cl	L			
Wentworth, James J	Pvt	M			
Wenzel, Julous L	Pvt	Hq			
West, Broady	Corp	A			
West, Mack	Corp	G			
West, Marvin E	Sgt	Hq			
West, Tenny B	Pvt	C			
West, Walter	Mech	F			

West, Walter	Pvt--1st Cl	I		K— Sept. 29-18	
Westbrook, Albert F	Pvt	M			
Whaite, Jesse G	Pvt--1st Cl	D	Dec. 6-18		
Whaley, Austin R	Pvt--1st Cl	L		K— Aug. 31-18	
Whedbee, John W	Pvt	A		S— Mar 16-19	
Wheeler, John J	Pvt	E			
Wheeler, Kelly B	Pvt	Inf			
Wheeless, Andrew J	Pvt--1st Cl	MG			
Wheeless, Billy	Pvt	B		W— Oct. 10-18	
Whichard, Julian D	Sgt	A			
Whichard, Marcus H	Sgt	K		W— Sept. 20-18	Dec. 2-18.
Whistnant, David C	Pvt--1st Cl	C			
Whistnant, John W	Wag	Sup			
White, Archie F	Corp	F		W— Sept. 29-18	Oct. 27-18.
White, Chester B	Corp	D		W— Sept. 29-18	Dec. 20-18.
White, Ernest N	Pvt	C		W— Aug. 31-18	
White, Gilbert B	Sgt	F		K— Sept. 29-18	
White, Granville D	Pvt	MG			
White, Jesse T	Pvt	D	Dec. 6-18		
White, John T	Pvt	B		W— Oct. 17-18	Dec. 2-18.

White, Leo A	Sgt	E			
White, Lyman	Corp	H			
White, McKenley	Pvt	D			
White, R obert E	Corp	MG			
White, Sam	Pvt	E		W— Oct. 10-18	
White, Thomas V	Corp	K			
White, William E	Pvt--1st Cl	K		D— Nov. 14-18	
Whitley, Harvey L	Corp	D			
Whitley, James A	Pvt--1st Cl	L			
Whitley, Walter C	Pvt--1st Cl	K		W— Aug. 27-18	
Whittier, Frank	Pvt	L		M— Aug. 31-18	Jan. 7-19.
Whitlock, Lemuel G	Pvt	D		T— Jul 15-18	
Whitner, Marion C	Pvt--1st Cl	D		K— Oct. 18-18	
Wickler, John D	Pvt	C		T— Jul 1-18	
Wickham, Frank W	Wag	Sup	Dec. 6-18		
Widger, Roy S	Pvt	MG		W— Sept. 29-18	
Wilenga, Nick	Pvt--1st Cl	B			
Wigand, Franklin P	Pvt	L		W— Oct. 17-18	Nov. 7-18.
Wiggins, Henry	Pvt	Sup			

Wiggs, Ed	Wag	Sup			
Wiggs, Richard P	Ck	C			
Wikle, Person	Pvt	F			
Wilbourn, Gibbs L	Corp	C			
Wilbourn, Johnnie	Pvt--1st Cl	I		K— Aug. 8-18	
Wild, Tate	Pvt	G		S— Sept. 1-18	Nov. 2-18.
Wilkerson, Ernest L	Corp	C		K— Oct. 17-18	
Wilkins, Charles R	Corp	F		W— Aug. 20-18 W— Oct. 10-18	
Wiley, James L	Pvt	C			
Wilkerson, Benjamin H	Sgt	G	Nov. 27-18		
Wilkerson, George W	Pvt	K		S— Aug. 9-18	Dec. 14-18.
Wilkerson, Harvey C	Corp	Hq		W— Sept. 29-18	Nov. 11-18.
Wilkerson, Raney P	Ck	Sup			
Wilkerson, Lucas	Pvt	H	Nov. 27-18		
Wilkins, Harvey N	Pvt	MG			
Wilkins, Junius J	Corp	C			
Willette, Leo W	Pvt	H		S— Oct. 29-18	
Williams, Albert W	Pvt--1st Cl	M		W— Oct. 10-18	Nov. 14-19.
Williams, Albert J	Pvt	A		W— Sept. 29-18	

Williams, Ambrose R	Sgt	H			
Williams, Arthur L	Sgt	Hq			
Williams, Asley C	Cpl	L		T— Feb. 8-19	
Williams, Boyd V	Sgt	Hq			
Williams, Claud	Pvt--1st Cl	K			
Williams, Daniel W	Corp	I		W— Aug. 30-18 W— Aug. 31-18	Nov. 11-18.
Williams, Frank A	Corp	K		S— Mar 16-19	
Williams, Frank W	Pvt--1st Cl	K			
Williams, George D	Sgt	A		W— Sept. 29-18	
Williams, Gomer	Sgt	I		T— Oct. 1-18	Jan. 3-19.
Williams, Henry D	Corp	M			
Williams, Julius L	Ck	K			
Williams, Jack	Pvt--1st Cl	K		K— Sept. 29-18	
Williams, John R	Corp	H		W— Aug. 1-18	
Williams, Lattie O	Sgt	B			
Williams, Lee B	Pvt	D		W— Oct. 17-18	
Williams, Lucian W	Pvt	MG		S— Nov. 29-18	
Williams, Roy	Corp	I		W— Sept. 29-18	
Williams, Sam W	Pvt--1st Cl	F		W— Sept. 29-18	Dec. 19-18.

Williams, Sam	Pvt	E		W— Sept. 28-18	
Williams, Sidney A	Pvt--1st Cl	A		W— Jul 21-18	
Williams, Taylor	Corp	I			
Williams, Thomas B	Pvt--1st Cl	Inf			
Williams, Taylor W	Pvt	D		W— Oct. 10-18	
Williams, Voyle J	Pvt--1st Cl	A			
Williams, William A	Corp	D		W— Sept. 29-18	
Williams, Wesley W	Pvt	K			
Williamson, Arthur E	Corp	D		W— Oct. 17-18	Nov. 11-18.
Williamson, Ellis W	Sgt	Inf		W— Sept. 27-18	
Williamson, Phillip A	Corp	E		W— Aug. 21-18	Sept. 15-18.
Williamson, Fletcher A	Pvt	E		W— Sept. 29-18	Nov. 11-18.
Williamson, Robert	Pvt--1st Cl	L			
Williford, Richard D	Pvt--1st Cl	MG		W— Sept. 29-18	Oct. 22-18.
Willis, David R	Pvt	C			
Willis, Weldon	Pvt--1st Cl	E		W— Oct. 17-18	
Willoughly, Ernest	Pvt	D		W— Sept. 29-18	
Willoughby, Ezra H	Pvt	C			
Wilmoth, Andy C	Mech	G		W— Sept. 29-18	Nov. 26-18.
Wilson,	Cpl	D			

Berton J					
Wilson, Claudie C	Pvt	F		W— Sept. 29-18	Dec. 19-18.
Wilson, Elias M	Pvt	Inf		T— Jan. 3-19	
Wilson, Franklin E	Pvt	G		D— Oct. 16-18	
Wilson, Fred	Pvt	G		W— Sept. 29-18	
Wilson, Grover O	Pvt--1st Cl	A		W— Sept. 29-18	
Wilson, Harrison A	Pvt	L			
Wilson, Herbert C	Ck	G			
Wilson, Isaac G	Pvt--1st Cl	K			
Wilson, James A	Sgt	G		W— Sept. 29-18	Dec. 2-18.
Wilson, Jarvis	Pvt--1st Cl	Inf			
Wilson, John A	Pvt	C		W— Oct. 17-18	
Wilson, John T	Pvt	L		W— Aug. 31-18	Sept. 1-18.
Wilson, John R	Cpl	L		K— Aug. 31-18	
Wilson, John S	Pvt	M	Dec. 6-18		
Wilson, Lanthan A	Sgt	MG		W— Sept. 29-18	
Wilson, Lube T	Ck	K			
Wilson, Michael E	Pvt	Hq	Dec. 6-18		
Wilson, Plato S	Pvt	C		W— Oct. 17-18	
Wilson, Rollie C	Pvt	D			

Wilson, Samuel	Pvt	F	Oct. 26-18		
Wilson, Walter W	Pvt	D		S— Mar 16-19	
Wineberger, Arthur C	Pvt	A			
Winfrey, Isaac E	Pvt	C		W— Aug. 31-18	
Winkler, George	Pvt	H		W— Sept. 29-18	Nov. 26-18.
Winkler, Thomas A	Pvt--1st Cl	MG			
Winner, Clyde W	Pvt	Hq	Feb. 20-19		
Winningham, James E	Cpl	B		W— Jul 26-18	
Winslow, Clenny	Pvt	D		D of W— Oct. 10-18	
Winslow, Grover L	Cpl	K		W— Oct. 10-18	
Winstead, Samuel	Pvt--1st Cl	F		W— Sept. 29-18	Dec. 19-18.
Winners, Charlie	Pvt	F		S— Aug. 14-18	
Wize, Thomas M	Pvt--1st Cl	A			
Wiswall, Herbert	Pvt	MG			
Witherspoon, John H	Pvt	D		T— Jan. 2-19	
Wlazlak, Mike	Pvt	D			
Wojtasiak, Jose	Pvt	M	Dec. 6-18		
Wold, Daniel C	Wag	Hq			
Wold, John I	Pvt--1st Cl	B		T— Feb. 4-19	
Wolfe, Albert D	Pvt	MG	Oct. 26-18		

Wolff, Richard R	Pvt	G		T— Sept. 12-18	
Womble, Erick G	Pvt	A		W— Oct. 17-18	
Wood, Francis H	Pvt--1st Cl	L		W— Oct. 10-18	Dec. 2-18.
Wood, Jesse	Pvt--1st Cl	C		W— Jul 18-18	Sept. 14-18.
Wood, John A	Pvt--1st Cl	F			
Wood, Leon C	Pvt--1st Cl	K		T— Sept. 2-18	
Woodward, Clarence C	Cpl	L	Oct. 26-18		
Woodard, Crobwell	Pvt	Sup		S— Nov. 11-18	
Woodard, Issac	Pvt--1st Cl	MG			
Woodard, Jacob	Ck	MG		T— Jan. 29-19	
Woodard, Roy L	Cpl	D			
Woodcock, Carl L	Pvt	H		S— Dec. 29-18	
Woodoll, Pertey	Pvt	MG			
Woodhall, James P	Pvt--1st Cl	Hq			
Woody, Ed H	Wag	Sup			
Woody, Jacob	Pvt--1st Cl	D		T— Jul 1-18	
Wooding, Charles S	Wag. Sup	Sup	Dec. 6-18		
Woodlief, William E	Pvt	B			
Woodliff, Joseph J	Ck	L		T— Jan. 29-19	

Woodring, Laud C	Pvt	MG			
Woodruff, William R	Wag	Sup			
Woods, Albert H	Pvt	I			
Woods, Charles H	Pvt	MG!	Oct. 26-18		
Woods, Frank G	Pvt--1st Cl	G		K— Sept. 29-18	
Woodward, Morgan T	Sgt	MG			
Woodward, Preston	Pvt	D		K— Oct. 10-18	
Woodward, Walter L	Pvt--1st Cl	K			
Woody, Charlie	Pvt--1st Cl	F			
Woolard, Claudius M	Cpl	I			
Woolridge, John	Pvt	C			
Woosley, Roy P	Pvt	H			
Wooten, Arthur L	Pvt	F	Dec. 6-19		
Wooten, Joseph S	Cpl	E		W— Oct. 10-18	
Wooten, Julian L	Cpl	K		W— Sept. 29-18	
Wooten, William H	Pvt	G		S— Aug. 27-18	Nov. 20-18.
Worley, Ambrose N	Cpl	I			
Warley, Wade A	Pvt	C		T— Jul 15-18	
Worrall, Charlie	Pvt	D		W— Oct. 17-18	

Worthington, Milton A	Pvt	D		W— Sept. 29-18	
Worthington, Ralph	Pvt--1st Cl	D			
Wrmy, Joseph	Pvt	M	Dec. 6-18		
Wright, Berkley D	Cpl	B		W— Sept. 29-18	Dec. 7-18.
Wright, George T	Pvt--1st Cl	B		T— Feb. 4-19	
Wright, James	Pvt--1st Cl	G		W— Oct. 10-18	Dec. 2-18.
Wright, James H	Pvt--1st Cl	I		W— Oct. 10-18	Dec. 14-18.
Wright, Simon P	Pvt--1st Cl	C		W— Jul 18-18	Dec. 2-18.
Wright, Thomas	Pvt	MG			
Wright, William A	Sgt	B		T— Feb. 4-19	
Wrinkle, Estel E	Pvt	Hq			
Wyatt, Moses L	Pvt	Hq		S— Dec. 7-18	
Wynn, Anderson	Sgt	L			
Wynne, Dee	Cpl	A		W— Sept. 3-18	Dec. 10-18.
Wynne, George B	Pvt	Hq		T— Mar 16-19	
Wynne, William B	Pvt	E			
Wynne, Hugh T	Cpl	Hq		T— Oct. 10-18	
Yates, Burns B	Pvt	H	Nov. 27-18		
Yates, Clarence	Pvt	H		W— Sept. 29-18	Dec. 8-18.

Yancey, Crawford G	Pvt	I			
Yearwood, Guy F	Pvt	D	Dec. 6-18		
Yestenes, Nicholini	Pvt--1st Cl	B		W— Oct. 17-18	Dec. 7-18.
Yock, Fred L	Pvt	I		W— Aug. 2-18	
Yonce, Osborne	Pvt	I			
Yopp, Samuel F	Sgt	MD			
Young, Bert G	Pvt	F		S— June 22-18	
Young, Boston A	Sgt	H		W— Sept. 29-18	Nov. 26-18.
Young, Ed	Pvt	F		M— Sept. 29-18	
Young, James R	BSM	Hq			
Young, James L	Pvt	G		W— Jul 26-18	
Young, Ralph A	Pvt	K			
Young, Rufus O	Pvt--1st Cl	C		S--May 1-18	
Young, Vincent	Ck	Sup			
Younger, Frank L	Sgt	C			
Yow, William M	Pvt	I		S— Nov. 22 18	
Zachary, Ora O	Pvt	I		W— Sept. 29-18	Dec. 5-18.
Zarkowski, William	Pvt	M	Dec. 6-18		
Zelker, George	Wag	Sup	Dec. 6-18		

Zabatka, Adam	Pvt	A		W— Aug. 29-18	
Zifchock, Martin A	Pvt	I		W— Sept. 29-18	Nov. 20-18.
Zimmerle, Albert R	Pvt	A	Dec. 6-18		
Zimmermann, Harry H	Pvt	Hq	Dec. 6-18		
Zimmermann, Larry	1st Sgt	K			
Zissels, Isadore	Pvt		Dec. 6-18		

LETTERS AND EXTRACTS

HEADQUARTERS 30TH DIVISION. AMERICAN EXPEDITIONARY FORCES.

March 11, 1919, France.

From: Major-General E. M. Lewis, Commanding 30th Div., A. E. F.

To: The Commanding Officer, 119th Infantry.

Subject: Service of the 119th Infantry.

1. While I have endeavored to express verbally to the officers my appreciation of their loyalty and service during the operations of the Division, I want you to take back with you this further testimony of my very great appreciation of the splendid record of the Regiment, as a whole, in the operations of the 30th Division. What has been accomplished by the Division is well known, and the 119th Infantry has had its full share in the accomplishment and can look back upon its experiences with nothing but pride, not only by its efficiency in the field, but also by its excellent discipline and military conduct, the 119th has always reflected credit upon the Division.

2. May all good things await you as individuals in just recompense for the great service done by you as an organization.

E. M. LEWIS,
Major-General, U. S. A.

JEFFERSON HOTEL,
Columbia, S. C., March 5, 1919.

DEAR METTS:

I cannot finally sever our close association during the past year without expressing to you my appreciation of the splendid efficiency of your command of the 119th Infantry, both in battle and in the trying period following the Armistice, when it sustained its high standard by attaining a high degree of perfection in drill, ceremonies, training and equipment.

I have never seen a finer regiment than the 119th Infantry when reviewed by General Pershing. The results attained were, in my estimation, due to your excellent judgment concerning the selection of your staff and to your refusal to retain any officer, regardless of personal feeling, unless he delivered the goods. Also to your happy faculty of inspiring your officers with real affection and securing absolute team work.

The spirit of your Regiment was always of the highest. Your cheerful attitude and spirit of enthusiasm and co-operation has ever been a satisfaction to us at Division Headquarters. You may well be proud of having developed one of the very best Regiments in the A. E. F.

With very best wishes for your success,

Cordially yours,

JOHN K. HERR,
Col. Chief of Staff, 30th Division.

HEADQUARTERS AMERICAN EXPEDITIONARY FORCES.
G-2-E--Visitors' Bureau

Paris, 37 Rue Bassano, February 12, 1919.

MY DEAR COLONEL:

I understand that you are about to return to the United States soon with your Division. As you of course know, I have not had the pleasure of serving with you when you were in actual command of your Regiment, beyond spending a couple of days in your camp just as you were leaving in December, 1917. I did, however, have the privilege of commanding the 119th Infantry for several months during your temporary absence, and I can but believe that the excellent spirit and morale which existed in that Regiment was largely the result of your work. I have seen personally no regiment formed during the war where officers and men gave a better or more loyal response under rather difficult conditions of training: Of the Regiment's work on this side I know only through hearsay, but comment that I have heard from outside the Regiment and Division has been most favorable.

Very sincerely yours,

F. LEJ. PARKER,
Brig. Gen. U. S. A.

COL. JOHN VAN B. METTS,
119th Infantry, A. E. F.
APC 749.

October 2, 1919.

From a letter written by Col. S. L. Faison, formerly Commanding General, 60th Brigade:

4. "This Regiment became one of the most effective fighting units in the Division, and, in my judgment, one of the best regiments in France."

FROM MAJOR-GENERAL E. M. LEWIS TO COL. METTS, MARCH 24, 1920.

"The 119th Infantry was an unusually efficient infantry unit, which did its full share in the operations of the 30th Division, under my command."

TOWNS AND VILLAGES CAPTURED BY THE 119TH INFANTRY.

Voormezeele.

Busigny.

Escaufourt.

St. Benin.

St. Souplet.

Ribeauville.

Ecaillon.

Mazinghiem.

Combined Operations Map
30th Div. A. E. F.
St. Quentin -- Cambrai Front
Sep 27 to Oct. 20
1918.

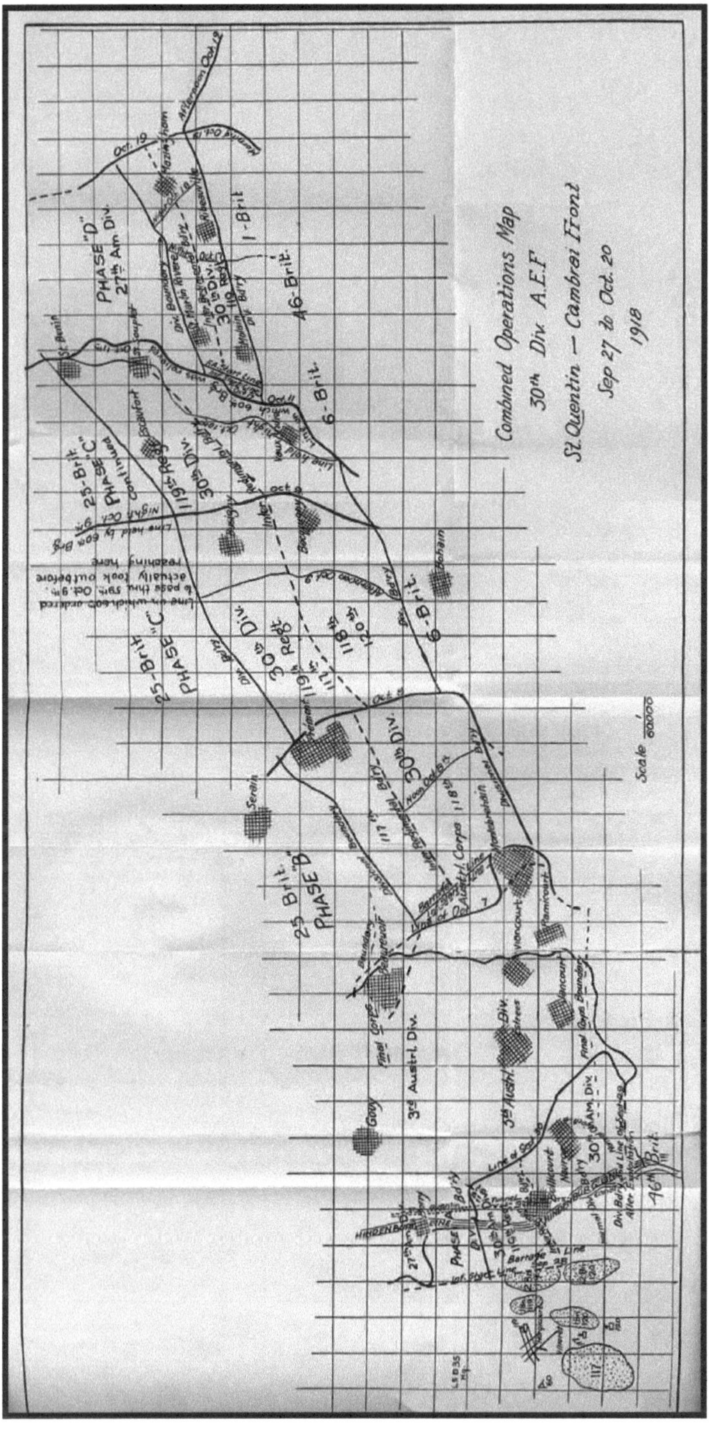
Combined Operations Map
30th Div. A.E.F
St. Quentin — Cambrai Front
Sep 27 to Oct. 20
1918